Fetal Heart Rate Monitoring
A Practical Guide

Ingemar Ingemarsson
Eva Ingemarsson
Department of Obstetrics and Gynaecology
University of Lund

and

John A. D. Spencer
Department of Obstetrics and Gynaecology
University College and Middlesex School of Medicine

Oxford New York Tokyo
OXFORD UNIVERSITY PRESS
1993

Oxford University Press, Walton Street, Oxford OX2 6DP
Oxford   New York   Toronto
Delhi   Bombay   Calcutta   Madras   Karachi
Kuala Lumpur   Singapore   Hong Kong   Tokyo
Nairobi   Dar es Salaam   Cape Town
Melbourne   Auckland   Madrid
and associated companies in
Berlin   Ibadan

Oxford is a trade mark of Oxford University Press

Published in the United States
by Oxford University Press Inc., New York

© Ingemar Ingemarsson, Eva Ingemarsson, and John A. D. Spencer, 1993

All rights reserved. No part of this publication may be
reproduced, stored in a retrieval system, or transmitted, in any
form or by any means, without the prior permission in writing of Oxford
University Press. Within the UK, exceptions are allowed in respect of any
fair dealing for the purpose of research or private study, or criticism or
review, as permitted under the Copyright, Designs and Patents Act, 1988, or
in the case of reprographic reproduction in accordance with the terms of
licences issued by the Copyright Licensing Agency. Enquiries concerning
reproduction outside those terms and in other countries should be sent to
the Rights Department, Oxford University Press, at the address above.

This book is sold subject to the condition that it shall not,
by way of trade or otherwise, be lent, re-sold, hired out, or otherwise
circulated without the publisher's prior consent in any form of binding
or cover other than that in which it is published and without a similar
condition including this condition being imposed
on the subsequent purchaser.

A catalogue record for this book is available from the British Library

Library of Congress Cataloging in Publication Data
Ingemarsson, I. (Ingemar)
[Elektronik fosterövervakning. English]
Fetal heart rate monitoring : a practical guide / Ingemar
Ingemarsson, Eva Ingemarsson, and John A. D. Spencer.
(Oxford medical publications)
Includes bibliographical references and index.
1. Fetal heart rate monitoring   I. Ingemarsson, Eva.
II. Spencer, John A. D.   III. Title.   IV. Series.
[DNLM: WQ 209 I463e 1993]
RG628.3.H42I5413   1993   618.3'207543—dc20   92-48525

ISBN 0-19-262269-2 (h/b)
ISBN 0-19-262268-4 (pbk.)

Typeset by Dobbie Typesetting Limited, Tavistock, Devon, UK
Printed in Great Britain by Dotesios Ltd
Trowbridge, Wiltshire

Foreword

Geoffrey Chamberlain, MD FRCS FRCOG,
Professor Obstetrics and Gynaecology,
University of London at St George's Hospital Medical School

The fetal heart rate (FHR) trace has been attacked by some for not being a valid measure of fetal well-being. This has arisen partly from a lack of understanding of what the trace can do; the lack of understanding of the basic science behind the trace has led many to over-diagnose and then be disappointed with the results of their own lack of logic. This has been reinforced by the judiciary using FHR traces in court against defendant obstetricians in claims for fetal damage. The law courts are not the best place to make scientific judgements, yet many have been passed against FHR measurements in legal decisions.

Continuous FHR monitoring probably started in the late 1950s and became widespread by the mid 1960s in America and Europe. It had a flourishing beginning in these epicentres following the work of Hon in America and Hammacher in Germany. By the time that fetal blood sampling came on the scene, pioneered by Saling in Berlin in the early 1960s, the FHR was a well-known and established method. Trying to persuade some obstetricians to use fetal blood sampling in the late 1960s and early 1970s was difficult for they had become fixed on FHR. Most of Europe stayed with the easier to perform FHR; America never liked fetal blood sampling partly because of the work involved and partly because of its invasiveness. Britain took them both, and those who used both methods widely did not consider that fetal blood sampling was superior to FHR but measured a different aspect of fetal behaviour. Single readings of scalp samples are not so helpful as a full reading of a fetal heart rate trace.

This book puts the FHR measurement into perspective. All the characteristics of the trace are considered and a total pattern of trace is shown to be predictive. Analysis of the full output is required for balanced judgement of the fetus, not just the timing of decelerations in relationship to nearby uterine contractions.

This volume sets out clearly the current knowledge of FHR; it explains the basic science of the subject in four chapters and then goes into detail about the parts of the trace, leaving the observer with a full sum of knowledge and ability to use the FHR safely and effectively. The inter-relationship with blood sampling is explained and justified. A full account of the clinical usage is given so that the reader will turn to the FHR with confidence. This is a book that is needed by all who work on labour wards and is fully recommended as a primer in the scientific interpretation of FHR traces.

Preface

Ascertainment of the presence of a fetal heart beat is no longer a sufficient indicator of fetal well-being during pregnancy and labour. Changes in fetal heart rate (FHR) correlate with neonatal outcome; this has led to the expectation that prospective FHR monitoring can be used to influence management to the benefit of mother and fetus. Technical advances in the mid-twentieth century have resulted in the ability to record the FHR continuously, and this has become an increasingly important aspect of the management of labour. Formal assessment of the value of continuous FHR monitoring in recent years has focused attention on the comparison between such records and 'old fashioned' auscultation of the FHR, without acknowledging that auscultation itself has never been evaluated impartially. Current practice remains confused by the conflict between the theoretical advantage of 'complete' FHR information, in the form of a continuous record, and data from controlled studies, which indicate increased rates of intervention without significant fetal benefit when compared with intermittent auscultation.

The basic physiological control of the FHR remains relatively poorly studied. Information derived from animal experiments indicates that the FHR falls in response to hypoxia. However, mechanisms that are believed to result in fetal hypoxia vary considerably in their threat to fetal well-being. Long-term fetal hypoxia results from chronic placental 'insufficiency' associated with proteinuric hypertension in the mother and is often well tolerated. Conversely, acute fetal hypoxia resulting from placental abruption is often fatal. Interruptions of uterine and umbilical blood flows, by a number of means, are often associated with transient reductions in FHR. The major difficulty remains our incomplete understanding of the mechanisms of the fetal response to these varying circumstances, and so it is often difficult to derive a full understanding of fetal well-being from the FHR alone. Thus, it is illogical to expect a particular FHR change, or pattern, to be diagnostic of a particular clinical problem. It is likely that similar FHR appearances occur as a result of the hypoxia itself, but appropriate management requires 'interpretation' of the clinical scenario to derive an understanding of the probable mechanism in each case.

It is our belief that appropriate use of FHR monitoring in labour is a safe and desirable practice. There can be no doubt that the experience of the last 30 years, particularly the increasing use of continuous FHR monitoring in uncomplicated labours, has contributed immensely to our understanding of FHR control.

This practical guide to FHR monitoring contains illustrations derived solely from continuous FHR monitoring records. This is necessary because it is not possible to illustrate auscultation, an auditory experience, by visual means. However, we are at pains to point out that the teaching of FHR 'patterns' and deviations from 'normal' during pregnancy and labour does not imply that continuous records should be used routinely. Intermittent FHR records, as well as auscultation of the FHR, seem perfectly appropriate in the context of normal labour progressing in the absence of clinical circumstances known to increase the risk of fetal hypoxia.

The objective of this book is to share our experience and knowledge about FHR monitoring in a visual form. This will help the reader to understand the advantages and disadvantages of this form of fetal evaluation. A number of questions remain unanswered because of limitations in our present state of knowledge. This book cannot, in any way, substitute for clinical experience and we continue to educate ourselves and our colleagues by weekly reviews of cases. However, we hope this book will assist and encourage those who wish to begin the process of understanding the FHR in clinical practice.

Lund   I.I.
London   E.I.
October 1992   J.A.D.S.

Contents

List of abbreviations		xiii
1.	Continuous or intermittent fetal heart rate monitoring?	1
	Introduction	1
	Advantages of continuous FHR monitoring	1
	Disadvantages of continuous FHR monitoring	4
	Side-effects	10
	Summary	10
2.	Technical aspects of fetal heart rate monitoring	12
	Introduction	12
	Signal loss	12
	Signal processing techniques	18
	Summary	24
3.	Maternal and fetal acid–base balance	27
	Introduction	27
	Fetal hypoxia	27
	Clinical interpretation of fetal scalp pH values	29
	Role of fetal scalp pH sampling	30
	Umbilical cord acid–base measurement	33
4.	Uterine activity	34
	Introduction	34
	Effect of contractions on intervillous perfusion	34
	Physiology of uterine activity	36
	Monitoring uterine activity	42
	Quantification of uterine activity	45
	Abnormal uterine activity	49
	Summary	49
5.	Control of fetal heart rate variability	53
	Introduction	53
	Autonomic nervous system influences	53

	Other influences on FHR	54
	FHR variability	54
	FHR reactivity and fetal behavioural states	67
	Interpretation of FHR variability	74
	Summary	82
6.	**Baseline fetal heart rate**	83
	Introduction	83
	Normal range	84
	Tachycardia	87
	Bradycardia	99
	Summary	110
7.	**Accelerations**	113
	Introduction	113
	Mechanism of FHR accelerations	113
	Clinical importance of accelerations	122
	Summary	124
8.	**Uniform decelerations**	125
	Introduction	125
	Classifications	125
	Characteristics	128
	The timing of uniform decelerations	130
	Summary	152
9.	**Variable decelerations**	153
	Introduction	153
	Aetiology	153
	Fetal oxygenation and acidaemia	157
	Predictive value	159
	Management	174
	Summary	180
10.	**Combined decelerations**	182
	Introduction	182
	Aetiology	183
	Summary	191
11.	**Prolonged decelerations**	193
	Introduction	193
	Aetiology	193

	Management	199
	Summary	202
12.	Second stage	203
	Introduction	203
	Clinical aspects	203
	Interpretation of the FHR pattern	206
	Summary	212
13.	Preterm fetal heart rate patterns	213
	Introduction	213
	Clinical aspects	213
	Factors influencing the FHR	213
	Summary	222
14.	Antenatal cardiotocograph	223
	Introduction	223
	Descriptive features	223
	Recording procedure and interpretation	232
	Clinical use of antepartum FHR records (non-stress tests)	240
	Technical problems	243
	Abnormal uterine activity	247
	The oxytocin challenge test (OCT)	250
	Summary	254
15.	Terminal fetal heart rate patterns	255
	Introduction	255
	Incidence of stillbirth	255
	Aetiology of stillbirths	256
	Terminal FHR patterns	259
	Summary	264
16.	Effects of drugs on the fetal heart rate pattern	265
	Local anaesthetics	265
	Oxytocin	267
	β-receptor agonists	267
	β-receptor blockers	268
	Atropine	269
	Narcotic sedatives, hypnotics, and analgesics	269
17.	Admission Test and fetal stimulation	274
	Introduction	274
	Principles of the Admission Test	274

	Interpretation of the Admission Test	275
	Use of the Admission Test	277
	Fetal stimulation in labour	281
18.	Clinical considerations	286
	Introduction	286
	Outcome measures	286
	Recommendations for monitoring in clinical practice	295
	Summary	301

Bibliography 302

Index 319

Abbreviations

ARM	artificial rupture of membranes
bpm	beats per minute
CTG	cardiotocograph (all records are 1 cm per min)
ECG	electrocardiograph
FHR	fetal heart rate
MU	Montevideo unit
NST	non-stress test
OA	occiput anterior
OCT	oxytocin challenge test
OP	occiput posterior
REM	rapid eye movements

1
Continuous or intermittent fetal heart rate monitoring?

Introduction

The technique of continuous fetal heart rate (FHR) monitoring has attained wide acceptance in clinical obstetric practice since its introduction during the late 1960s. However, although there have been considerable advances in the technical specifications of monitors, increasing experience in recent years, particularly in an increasing number of low-risk pregnancies, has shown that interpretation of fetal well-being from a continuous record of FHR has limitations. The concept of cost versus clinical benefit continues to be the subject of much debate and there is concern about the use of continuous FHR monitoring in clinical practice without adequate training of the staff who use it.

Advantages of continuous FHR monitoring

The production of a continuous, graphical record of the FHR was thought to have clear advantages over intermittent auscultation and therefore led to expectations of improvements in obstetric care. The apparent advantages are listed below.

A complete and accurate record of the FHR

Calculation of the FHR using a stethoscope can be difficult and the duration of auscultation is of great importance. If the value counted during 15 s is multiplied by four to obtain the rate per minute then there is potential for introducing a fourfold error (Fig. 1.1). As well as an accurate baseline, a continuous record of FHR allows FHR baseline variability to be seen. Although careful auscultation might reveal accelerations related to fetal movements, it is impossible to identify FHR variability or shallow decelerations (Figs 1.2 and 1.3).

A precise indication of the relationship with contractions

By plotting a continuous record of uterine activity on the same paper strip as the FHR, it is possible to see clearly any related changes in FHR, such as

2 Continuous or intermittent fetal heart rate monitoring

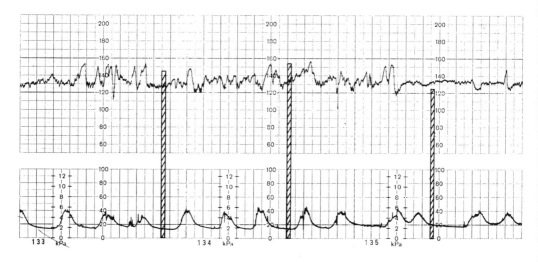

Fig. 1.1 Illustration of the different fetal heart rates obtained by intermittent auscultation and continuous FHR recording.

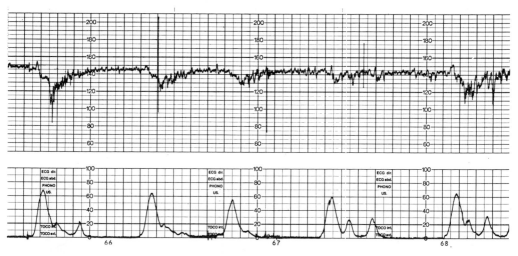

Fig. 1.2 A continuous record of intrauterine pressure and FHR showing a baseline FHR of 145 bpm with variability between 5 and 10 bpm amplitude. Discrete decelerations following contractions, which would not have been auscultated, are clearly evident.

accelerations and decelerations. The composite record is known as a cardiotocograph (CTG). A continuous record of contractions also enables the assessment of total uterine activity, particularly if an intrauterine pressure catheter is used. Basal tone between contractions and the pressure generated

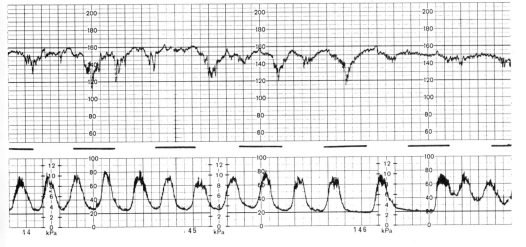

Fig. 1.3 Uterine activity, recorded by an external tocometer, showing abnormally frequent contractions. The baseline FHR is 155–160 bpm with low (less than 5 bpm amplitude) variability. Discrete late decelerations can clearly be seen after the peak of each contraction. Intermittent auscultation would have identified the tachycardia but may have missed the decelerations.

during each contraction cannot be measured without an intrauterine catheter but amplitude, duration, and frequency of contractions can be recorded by external tocography.

Permanent record of events

There need no longer be doubt about whether the fetal heart has been heard correctly. Such discussions were not uncommon in connection with intrauterine fetal deaths when a stethoscope had been used for monitoring the FHR. These tragic cases may now be partly explained by the fact that the fetus often maintains baseline FHR, even when it is severely hypoxic. Imminent fetal death cannot be identified by stethoscopic auscultation unless a bradycardia is present. Use of continuous CTG records has shown such cases to have subtle abnormalities, such as low or absent baseline variability and shallow, late, FHR decelerations (Fig. 1.4). Given this new means of surveillance, many high-risk pregnancies, who would otherwise have had an elective Caesarean section on the basis of a poor obstetric history (for example elderly primipara, infertility, or previous obstetric catastrophe) were allowed to labour.

Antenatal assessment of placental function

Antenatal records of FHR began to replace hormonal tests of placental function because it was possible to get an immediate 'result' by which to assess fetal

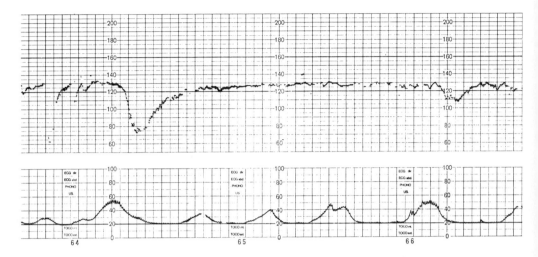

Fig. 1.4 External monitoring of uterine contractions and FHR. The baseline FHR is 130 bpm and variability is almost absent. Occasional late decelerations are present, which would probably have been missed by intermittent auscultation.

condition. The antenatal CTG has now become the most frequently used method of fetal surveillance in the third trimester of pregnancy.

Reassurance of normality

Enthusiasm for the new technique was often shared by parents, who were now able to hear the fetal heart beating and observe fetal activity. Many parents admitted a feeling of increased security with electronic FHR monitoring.

Disadvantages of continuous FHR monitoring

As the use of continuous FHR in clinical practice increased, it was not long before some disadvantages became apparent.

False-positive FHR changes

In low-risk pregnancies, FHR changes were found in 5–10 per cent of antenatal records, depending upon the method of classification. In labour, FHR changes were frequently seen in low-risk women.

A normal CTG record occurs in only 50 per cent of all labours (Table 1.1). About 15 per cent of all records have a baseline abnormality. With normal baseline ranging from 110 to 150 beats per minute (bpm) some 10 per cent of

Table 1.1 Incidence of CTG changes during labour in an unselected population

Baseline (%) (>40 min observation)	
normal	85.3
tachycardia (>160 bpm)	6.2
bradycardia (<120 bpm)	8.5
Baseline FHR variability (%) (>40 min observation)	
saltatory (>25 bpm)	4.6
normal	85.2
decreased (5–10 bpm)	7.7
silent pattern (<5 bpm)	2.5
Decelerations (%) (repetitive)	
none	54.0
variable	29.0
early	11.5
late	1.9
combined	2.6
undefined	1.0

cases will have a tachycardia while the frequency of bradycardia is less than 10 per cent. In some 15 per cent, changes in FHR variability occur; absent variability, however, is infrequent (2.5 per cent).

In half of all records in labour there are decelerations during the first stage of labour; the most common deceleration is described as 'variable' (Fig. 1.5).

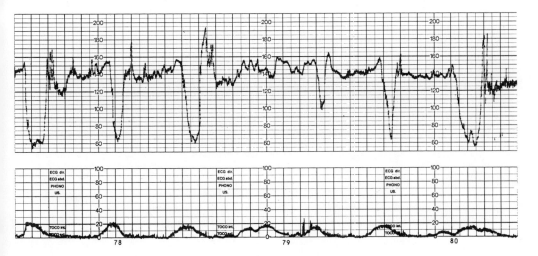

Fig. 1.5 An external CTG showing a baseline FHR of 140 bpm with normal variability and variable decelerations.

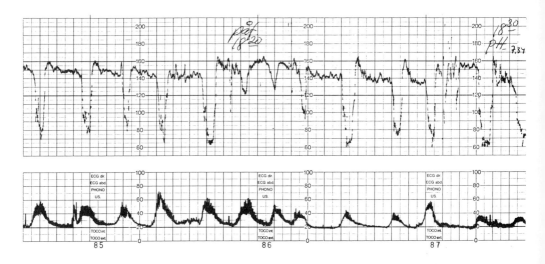

Fig. 1.6 An external CTG showing a baseline FHR of 140–150 bpm with normal variability. Typical variable decelerations are shown with rapid drops in FHR to 60–80 bpm lasting 30–45 s. The return to the baseline FHR after the peak of each contraction is also rapid. The presence of accelerations and the normality of the baseline FHR suggest that fetal hypoxia is unlikely. The fetal scalp blood pH was 7.34.

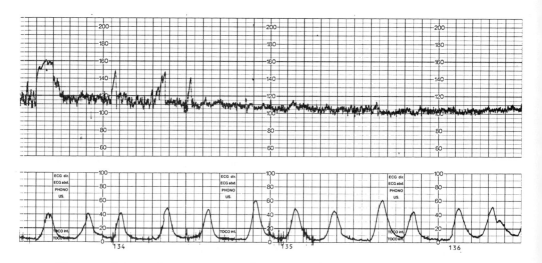

Fig. 1.7 Intrauterine pressure and FHR record showing a baseline FHR of 105–115 bpm with normal variability. There are no decelerations and accelerations indicate very little risk of fetal hypoxia despite the low FHR (baseline bradycardia).

However, such decelerations with a duration of more than 60 s and an amplitude of more than 60 bpm, or dropping to a rate of 70 bpm, or connected with other pathological features will only constitute a small percentage of all tracings.

Overreactions to all CTG changes are inevitable and lead to unnecessary interventions. Caesarean sections were done for suspected fetal distress when records showed mild and moderate variable decelerations or uncomplicated bradycardia (Figs 1.6 and 1.7).

Poor interpretation of CTG traces

Adequate knowledge of fetal physiology is important for appropriate interpretation of FHR changes. Initially, knowledge regarding FHR variability, fetal reactivity, and behavioural states was poor and the importance of FHR accelerations was not recognized. The value of continuous FHR monitoring for screening purposes was soon acknowledged such that the risk of delivering a hypoxic fetus in the presence of a reactive trace, without pathological changes in labour, is small (Figs 1.8 and 1.9). However, the reverse does not hold true, and pronounced CTG changes do not correlate well with fetal hypoxia and acidaemia. The experience of the last 20 years has shown that FHR changes in the first stage of labour must be interpreted with great care, as more than 90 per cent are innocuous.

However, 6–7 per cent of CTG changes in labour demand special attention. Half of these ominous CTG changes occur late in first stage of labour and will thus not offer any great clinical problems because delivery is imminent or may easily be expedited. The remaining 3–4 per cent will require expertise in FHR interpretation, with an adequate understanding of fetal acid–base balance and the development of hypoxia, if unnecessary intervention is to be avoided.

Increased Caesarean section rate for fetal distress

At about the time continuous FHR monitoring was introduced into clinical practice, the Caesarean section rate began to increase. However, this was not merely the result of overdiagnosis of fetal distress. Several other reasons for performing an abdominal delivery, such as for breech presentation, preterm labour, and prolonged labour, were becoming accepted practice. Continuous FHR monitoring has certainly contributed to an increased rate of Caesarean section in many departments, although some centres reported unchanged Caesarean section rates for fetal distress (1–2 per cent) after introduction of continuous monitoring. Out of 5000 low-risk patients delivered at the University Hospital, Lund, Sweden, during 1977 and 1978, only 30 patients (0.6 per cent) were delivered by Caesarean section for fetal distress.

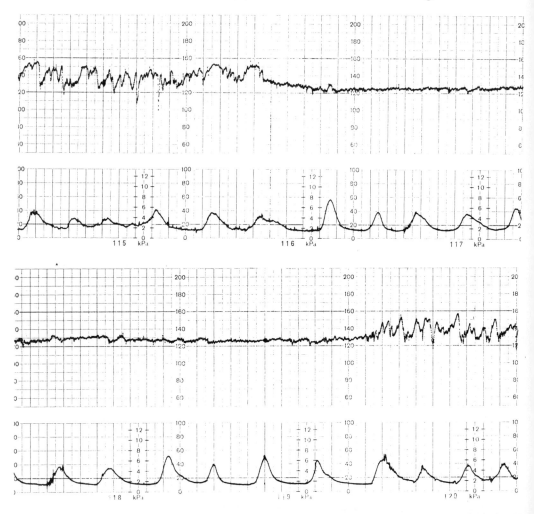

B

Fig. 1.8 An external CTG showing a baseline FHR of 120–130 bpm with an episode of low variability lasting 36 min. This is consistent with a period of fetal quiescence indicating normal fetal behavioural state changes during labour. Further investigation of such low variability is not required when evidence of fetal rest–activity cycles is present, indicated by a reactive FHR pattern before and afterwards.

Poor predictive value of FHR changes

The correlation between abnormal FHR changes and the condition of the baby at birth is poor. Even the so-called ominous trace has a poor correlation with fetal acidaemia or a low Apgar score at birth. It may seem incongruous that

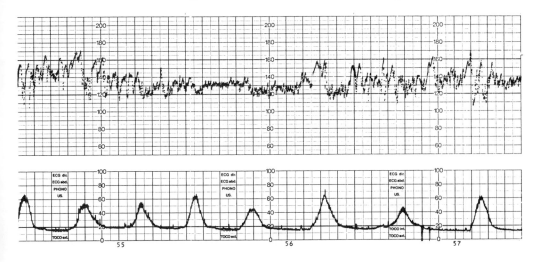

Fig. 1.9 An external CTG showing a normal FHR pattern.

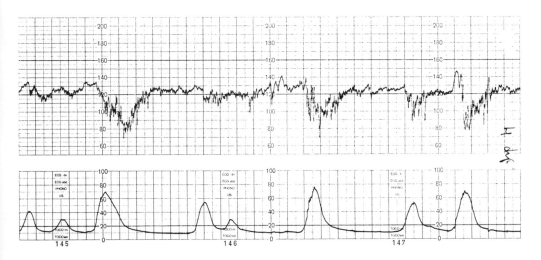

Fig. 1.10 Intrauterine pressure and FHR showing repeated late decelerations indicating hypoxaemia associated with the reductions in placental circulation during contractions. Further investigation of such a pattern by fetal scalp blood pH sampling is indicated.

a baby may be born with good Apgar scores and a normal cord blood pH after demonstrating repetitive late decelerations in labour. However, late decelerations in such cases are often accompanied by normal baseline FHR and variability and fetal reactivity (Fig. 1.10). Experience has shown that such records do not relate well to fetal distress.

Several factors other than hypoxia may cause a depressed baby at birth and may also occur in the presence of a normal trace:

(1) trauma (operative interventions, shoulder dystocia);
(2) sedatives, analgesics, or hypnotics;
(3) infections (chorioamnionitis);
(4) meconium aspiration;
(5) congenital malformations;
(6) fetal anaemia;
(7) prematurity;
(8) technical problems.

Modern monitors can easily obtain good records of FHR and contractions. However, it is essential to make sure that the signal is from the fetus rather than mother, especially in cases of prematurity and FHR arrhythmia. If auscultation is unable to resolve this question then real-time ultrasound may be advisable.

Side-effects

To obtain an adequate record of the FHR it may be necessary to perform amniotomy. However, both early and variable decelerations are more frequent after amniotomy and this may increase concern about fetal well-being. Modern monitors are able to make good records of the FHR using external transducers, and intermittent monitoring is considered sufficient in most cases. Thus, continuous FHR monitoring needs to be used only in selected cases.

Summary

Continuous FHR monitoring is by far the most common method of surveillance for fetal well-being in pregnancy and labour. Approximately three-quarters of all deliveries in the USA and Western Europe are monitored in labour. After more than 20 years of experience it should be possible to use FHR monitors in a way that is acceptable from a medical point of view as well as meeting the wishes of consumers. A sensible and sensitive attitude by staff towards pregnant women who wish to avoid too much technological intervention is a prerequisite of good practice. There is no conflict between adequate monitoring of mother and fetus during pregnancy and labour and a warm and caring attitude by medical and midwifery staff. This includes tailoring the level of monitoring to the individual, taking into account risk factors, and combining accurate

Summary

monitoring with a low frequency of interventions. Experience with the use of cardiotocography means that intermittent auscultation can be performed with increased confidence in the absence of risk factors or complications.

All staff should have adequate knowledge of fetal physiology and be trained to interpret changes in FHR. The aim of this book is to bring together the basic principles of FHR monitoring, many of which are the result of experience with continuous records.

2
Technical aspects of fetal heart rate monitoring

Introduction

Continuous records of FHR and uterine activity can be obtained by a number of different techniques, although clinical practice relies on two methods of obtaining the FHR and one method of obtaining uterine contractions. The resulting paper record—the cardiotocograph (CTG)—is the final product of signal processing by the electronic monitor, which depends upon the method used for obtaining the signal. All monitors now have a microprocessor, which is programmed to enhance the true signal by identifying and discarding 'noise'. This 'logic' principle accepts heart beart signals corresponding to rates of 50–210 bpm (30–230 in the USA) provided that beat-to-beat variation is not outside certain limits. This often results in a delay of a few seconds before the chart record commences and also means that there will be no printout in the absence of an 'acceptable' signal.

Internal monitoring refers to direct contact with the fetus to obtain the ECG and is performed only during labour. External ultrasound is used antenatally and during labour with intact membranes to obtain the FHR via the mother's abdominal wall. Such signals may be of variable quality and are therefore electronically processed at the expense of beat-to-beat FHR variability. Technical quality (signal loss) depends upon the method used to obtain the signal.

Applicability of three methods (phonography, ultrasound, and abdominal ECG) was studied in 163 patients at the University Hospital, Lund, between weeks 34 and 40 of pregnancy. More than 90 per cent of patients had a record that was technically good with at least one of the three methods, but only 14 per cent had a good record with all three.

Signal loss

Signal loss is greater when external ultrasound is used to identify heart beats, and the printout will usually cease until acceptable signals are identified again. Signal loss occurs if excessive fetal movement interferes with the Doppler ultrasound used by the external transducer during pregnancy. Loss may also

Signal loss

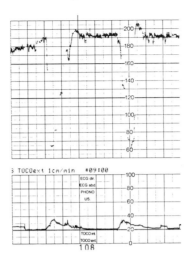

Fig. 2.1 Pronounced variable decelerations.

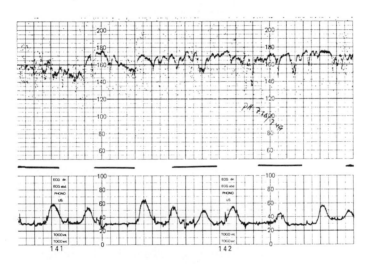

Fig. 2.2 A poor quality record by scalp electrode due to frequent fetal extra systoles. Fetal scalp blood pH was 7.36.

occur with large variable decelerations in labour if the FHR changes too quickly (Fig. 2.1). Frequent extrasystoles will also disturb the signal (Fig. 2.2) and an FHR arrhythmia may prevent a record being obtained (Fig. 2.3). Even with direct contact using a fetal scalp electrode to obtain the fetal ECG, records may have signal loss as a result of loss of contact or FHR arrhythmias (Figs 2.4 and 2.5).

Technical aspects of fetal heart rate monitoring

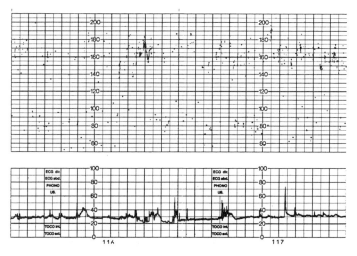

Fig. 2.3 This record is impossible to evaluate because of a fetal heart arrhythmia.

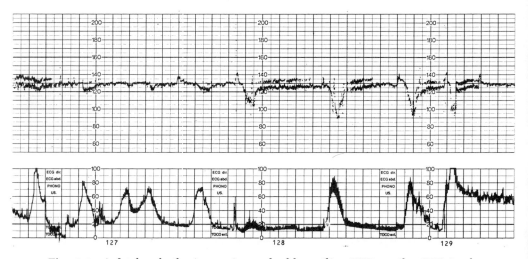

Fig. 2.4 A fetal arrhythmia causing a double outline FHR on the CTG (scalp electrode).

If the continuous FHR record is technically unsatisfactory (Fig. 2.6) then the FHR must be determined by auscultation using a Pinard stethoscope or Doppler ultrasound device. It may be necessary to use real-time ultrasound to confirm fetal heart movements, as with partial heart block, which will cause the FHR to vary greatly (e.g. from 80 to 160 bpm). This may resemble a CTG where maternal and fetal heart signals are recorded alternately.

Correct identification of the signal obtained by an external ultrasound transducer is essential. Administration of β-adrenergic receptor agonists to

Signal loss

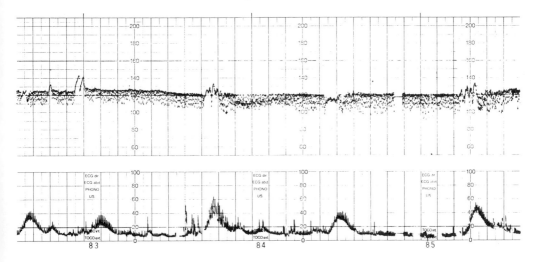

Fig. 2.5 The effect of a fetal arrhythmia on the FHR (scalp electrode).

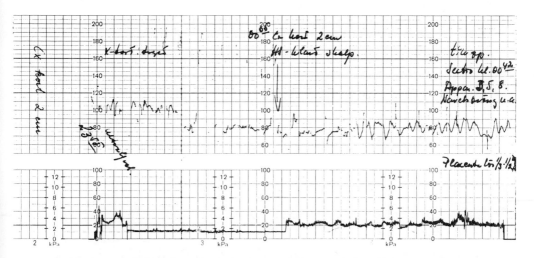

Fig. 2.6 A CTG performed on admission in labour (Admission Test) showing bradycardia and reduced FHR variability in a primipara at term. A fetal scalp electrode was attached after rupture of membranes (at 3) but problems with the FHR record continued due to tetanic uterine activity. A Caesarean section was performed because of a suspicion of placental abruption. Apgar scores were 3, 5, and 8.

inhibit preterm labour often results in a maternal tachycardia of 120–130 bpm. This may be identified from the pulsations of the large lower abdominal or pelvic arteries by the ultrasound beam and will produce a regular signal, which may be mistaken for that of the fetus, unless the maternal heart rate is noted

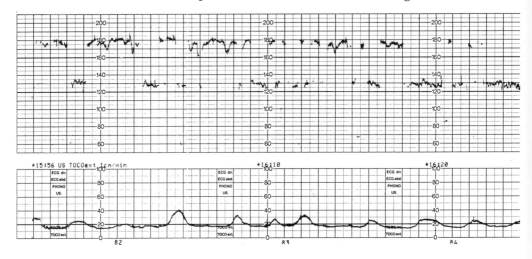

Fig. 2.7 A primipara at 31 weeks' gestation given an infusion of β-mimetics for inhibition of preterm labour. The CTG recorded both maternal heart rate (125–130 bpm) and fetal heart rate (175 bpm).

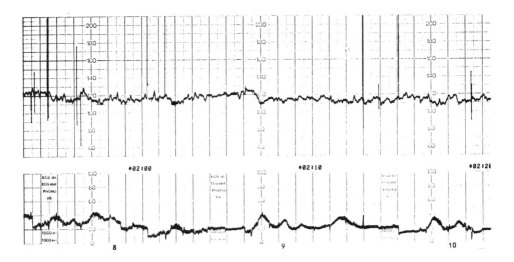

Fig. 2.8 A maternal heart rate record by scalp electrode from a dead fetus.

(Fig. 2.7). Similarly, a CTG of the maternal heart rate may be produced after fetal death, even using a direct fetal scalp electrode (Figs 2.8 and 2.9) by recording the maternal electrocardiograph (ECG). Decelerations and accelerations of the maternal heart rate related to contractions occur and a maternal heart rate exceeding 100–110 bpm is not uncommon in labour.

Signal loss

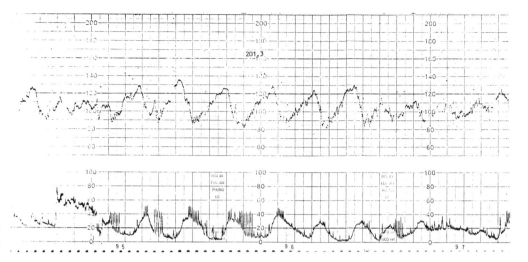

Fig. 2.9 A maternal heart rate record, by scalp electrode from a dead fetus, showing decelerations with contractions.

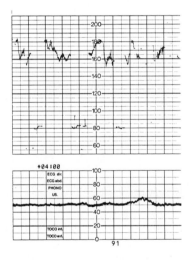

Fig. 2.10 Apparent halving of the FHR (160–165 bpm) by intermittent identification of the maternal heart rate.

Confirmation that the signal is not maternal should be performed routinely by comparison with the maternal heart rate at the onset of the record.

Another problem with external FHR monitoring is the risk of the signal being halved or doubled. Doubling may occur during a bradycardia when there is sufficient separation of the aortic and pulmonary valve movements from the atrioventricular valve for the ultrasound to identify these as separate heart beats; auscultation will clarify this. Halving of a normal rate (Fig. 2.10) will occur if alternate heart movements are missed by the ultrasound transducer, but the possibility of maternal signal should also be considered.

Signal processing techniques

Signal processing by the monitor varies according to the method used to obtain the FHR signal (ultrasound or ECG). The signal is amplified and filtered to reduce noise artefact and the fetal heart beat components require recognition. The original method of threshold detection identified all signals that crossed a preset level (in microvolts). Thus maternal and fetal ECG signals with amplitudes above this threshold were recognized, as were any noise artefacts of similar amplitude. Continuous ultrasound signals were also recognized by threshold detection of the ultrasound envelope. However, this often led to 'jitter' when the two components of each heart movement varied in amplitude such that recognition of different components gave rise to an artefactual short-term variability (Figs 2.11, 2.12 and 2.13). Tracings with such artefacts were often mistaken for normal variability, when in fact this was actually reduced. On the other hand, although reduced variability may result from excessive filtering of the signal or averaging of pulse intervals, absence of variability displayed on the trace is always likely to be a genuine absence of FHR variability. Modern monitors use autocorrelation to improve the signal-to-noise ratio (see below).

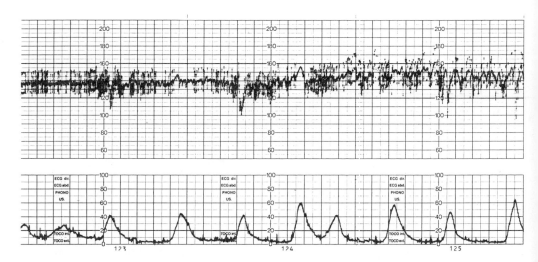

Fig. 2.11 'Gitter' causing a record with significant artefact. The variability is impossible to interpret.

Signal processing techniques   19

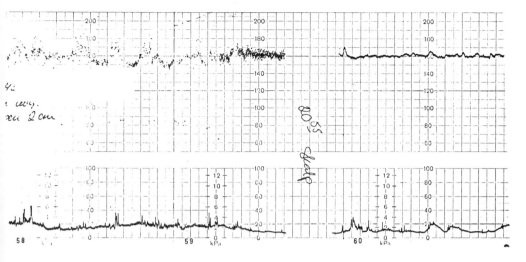

Fig. 2.12 An external (ultrasound) record simulating normal FHR variability. After application of a fetal scalp electrode the absence of variability was clearly revealed.

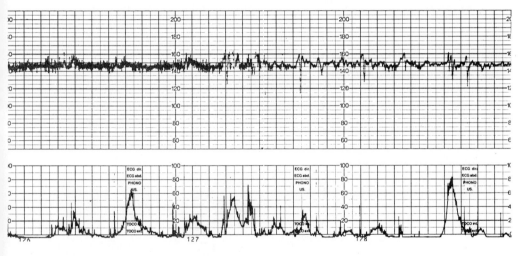

Fig. 2.13 An external ultrasound record of FHR. Initially, the trace contained much artefact but this disappeared after repositioning of the transducer over the fetal heart.

The signals are electronically processed to produce a continuous record without too many interruptions. Some averaging is performed and mean values are calculated from the arithmetic mean of three consecutive pulse intervals. An updated mean value is obtained with every new signal (moving average). Pulse intervals are transformed to rates and recorded as beats per minute (bpm). The averaging process results in loss of part of the pulse interval (beat-to-beat) variability. Accelerations and decelerations may be slightly transformed, but in technically acceptable recordings this is of little importance.

The record paper speed is critical when it comes to FHR pattern recognition. All records in this book are recorded at 1 cm/min, which is the most commonly used paper speed in Europe. Some centres use 2, 3 or even more cm/min to obtain a better display of FHR variability (Figs 2.14 and 2.15) — the drawback of course being cost and handling of paper.

Direct fetal electrocardiography

The ECG signal is obtained directly from the fetus using a scalp electrode, which is little influenced by maternal position. The electrode is a curved or spiral needle of stainless steel that fixes to the scalp or buttock by piercing the skin. An application device is necessary for some types. The design allows easy fixation and removal but is at the expense of a pin-prick puncture of the fetal skin. The presentation and position of the baby must be determined prior to electrode attachment to avoid trauma to its face or genital organs. The cervix needs to be dilated 0.5–1 cm to avoid trauma to maternal tissue. Care should be taken to avoid attachment of the electrode to the cervix (Fig. 2.16) or fetal membranes.

Internal monitoring during labour requires amniotomy. Such intervention should not be undertaken without good reason, such as a need for continuous monitoring when an unsatisfactory signal is obtained by external means.

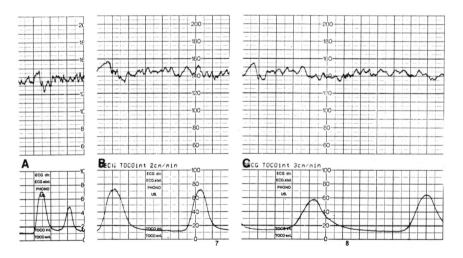

Fig. 2.14 An intrauterine record of uterine activity and FHR. Three 4-min periods are displayed at different paper speeds. A, 1 cm/min; B, 2 cm/min; C, 3 cm/min.

Abdominal electrocardiography

The amplitude of the fetal R wave of the ECG complex is much smaller when obtained from electrodes on the maternal abdomen. To eliminate confusion between fetal and maternal ECG complexes the fetal R wave was ignored if produced within a 'window' of 190 ms before and 75 ms after a maternal R wave (Hewlett Packard 8030 A) (Fig. 2.17). A fetal beat was substituted at an appropriate time interval between the preceding and the subsequent fetal ECG complex. In some records as much as 30 per cent of all signals were substituted. Studies showed that short-term (pulse interval) as well as long-term (baseline) FHR variability were reasonably well recorded. This method of monitoring was difficult and has become largely replaced by ultrasound.

Ultrasound FHR monitoring

Ultrasound is the usual technique for external monitoring. It is based on the Doppler shift of ultrasound that occurs after insonation of a moving object. The Doppler-shifted frequencies are in the audible range and relate to the velocity of movement. Clear, sharp sounds are generated by the fetal heart valve movements and the maximum frequency envelope is used to identify each heart beat. Pulse intervals are converted to a rate. Arterial blood flow, maternal movements and fetal movements may interfere with the signal.

Most monitors now use pulsed ultrasound generated from the abdominal transducer. This results in a wider sensitive window than that obtained when continuous ultrasound was used to identify the moving fetal heart. Pulsed Doppler ultrasound produces FHR records that compare well with records obtained by scalp electrode (Fig. 2.18). Good records are obtained from earlier gestations in pregnancy than were possible with continuous wave, non-autocorrelated techniques. Similarly, high quality external records can be obtained in labour, even in the second stage. This technique has reduced the need for amniotomy for application of a scalp electrode. The wider ultrasound beam means that more extraneous movements are identified and the success of pulsed ultrasound was the concurrent introduction of signal autocorrelation.

Modern monitors contain microprocessors, which contain algorithms for the signal to be processed by autocorrelation. The monitor compares the incoming signal with a stored version of the preceding one. Any signal with a regular component will generate a good autocorrelation function because the similar components of the signal will match well when correlated with each other. The intervals between peaks of this autocorrelation function actually reflect the rate of the regular component of the signal and these are used to derive the record. Provided the ultrasound transducer is directed at the fetal heart then the signal

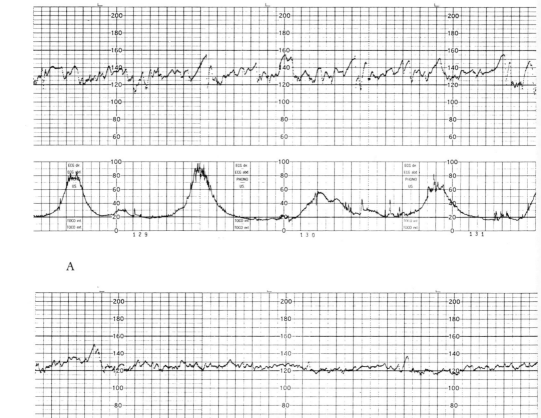

A

B

Fig. 2.15 (caption opposite).

will be the fetal heart rate. If the transducer moves it is possible for the monitor to autocorrelate the regular signal derived from the pulsations of blood flow in the large maternal vessels of the pelvis. Thus, rarely, a maternal heart rate may readily be generated by autocorrelated ultrasound, as described above (Fig. 2.19).

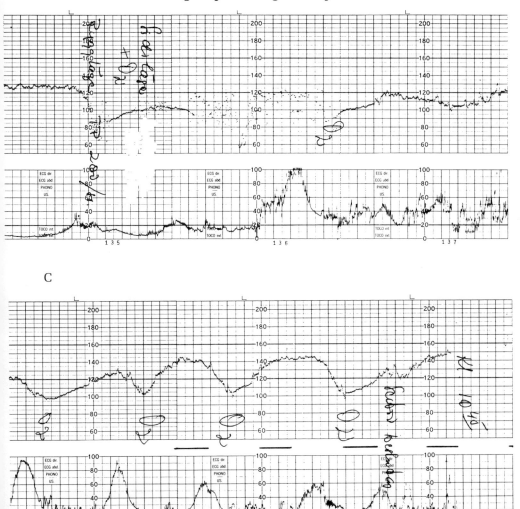

Fig. 2.15 A 1-h CTG during labour recorded at 2 cm/min. Uterine activity was recorded by external tocography. The woman had had a previous Caesarean section for fetal distress and had now been in labour at term for 2 h. A and B, Normal, reactive pattern; C, abrupt change in the FHR pattern occurred. Bradycardia, decreased variability, and pronounced late decelerations can be seen. Although vaginal bleeding was not evident, an emergency Caesarean section revealed a placental abruption. Apgar scores were 3, 5, and 10. There was no indication for fetal scalp blood sampling after such an unexplained ominous change in the CTG.

24 Technical aspects of fetal heart rate monitoring

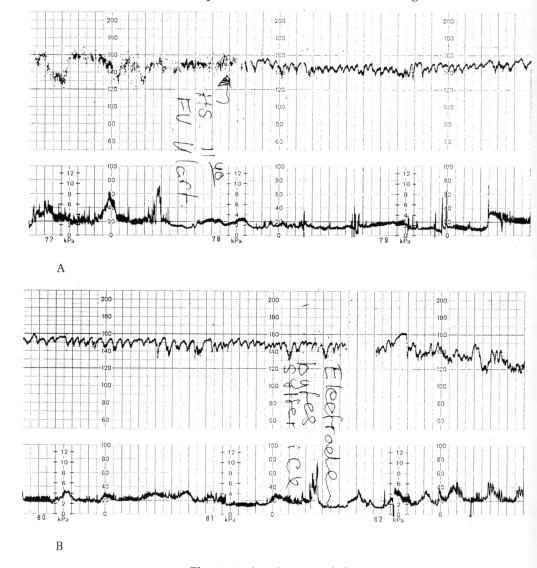

Fig. 2.16 (caption opposite).

Summary

A basic knowledge of how the FHR is derived from the signals obtained by fetal monitors is important. An understanding of the limitations of the signal processing methods will prevent the deduction of erroneous conclusions from the CTG record.

Summary

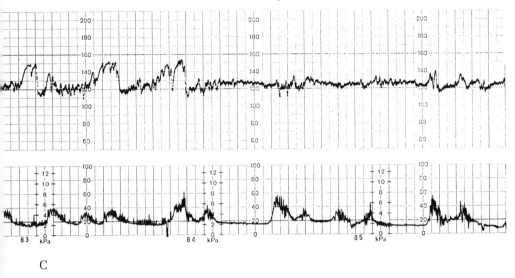

C

Fig. 2.16 An external record of uterine activity. A, A scalp electrode was applied at 78. The FHR pattern appeared as pseudosinusoidal. The scalp electrode was found attached to cervix; B and C, when the electrode was replaced on the fetal head (at 82) a normal FHR pattern was seen.

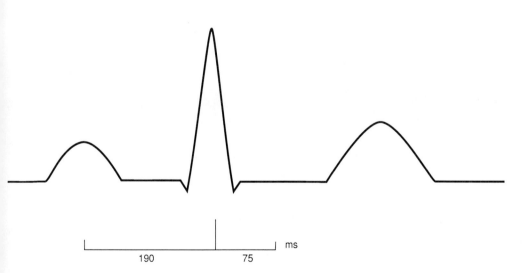

Fig. 2.17 A schematic picture of an ECG complex.

26 Technical aspects of fetal heart rate monitoring

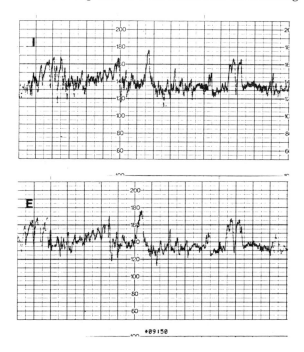

Fig. 2.18 Simultaneous records of FHR with a scalp electrode (I) and external ultrasound with autocorrelation (E).

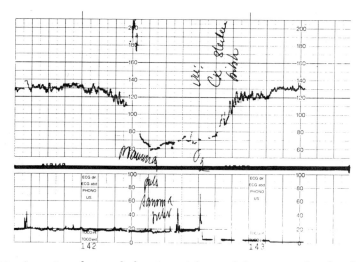

Fig. 2.19 An external record showing pick-up of the maternal pulse during a contraction. Note the apparent fall in FHR as the fetal signal is lost and the apparent rise again as the fetal signal is regained. However, discontinuity between fetal and maternal records provides the clue to interpretation as separate sources.

3
Maternal and fetal acid–base balance

Introduction

Fetal oxygen supply depends upon maternal ventilation, uterine perfusion, and adequate placental and fetal circulations. The oxygen tension in maternal arterial blood reaching the intervillous space is high and fetal blood, reaching the villi through the umbilical arteries, has a low oxygen tension. Umbilical venous blood, returning directly to the fetal heart, carries oxygen from the placenta for distribution to the fetal tissues. The placenta also uses oxygen for its metabolic processes and so umbilical venous oxygen tension never reaches the same level as uterine venous oxygen tension.

Fetal hypoxia

Two main factors contribute to the adequacy of fetal oxygenation despite the low oxygen tension. First, fetal haemoglobin is present in higher concentrations than in the adult and has a greater affinity for oxygen. Second, the fetus normally supplies more oxygen to its tissues than is required, with the result that the fetus is relatively resistant to mild/moderate hypoxia. Experimental data indicate that fetal oxygen supply can be reduced up to 50 per cent without apparent influence on the fetus.

A reduction in uteroplacental perfusion results initially in a greater extraction of oxygen by the fetus with a consequent reduction in the oxygen concentration in the uterine veins. Subsequently, there is a reduction in umbilical venous oxygen concentration but fetal hypoxaemia does not produce tissue hypoxia as long as the oxygen supply exceeds the needs of fetal tissues for aerobic metabolism. Ultimately, fetal tissue hypoxia occurs and the fetus changes to anaerobic metabolism with a build-up of lactic acid.

The fetus may become hypoxic in a number of ways. Occlusion of the umbilical vein occurs during uterine contractions, especially after rupture of the membranes. Hypertonic uterine activity reduces the maternal arterial input into the intervillous space more than normal and maternal hypotension, for example after an epidural, may also reduce intervillous space perfusion. Reduced fetal transport of oxygen occurs with anaemia (as in rhesus disease) and a chronic

Fetal response to hypoxia

The circulatory fetal response to hypoxia is a redistribution of cardiac output favouring the heart, adrenals, and brain. Cardiac output is the product of heart rate and stroke volume. Fetal stroke volume is near maximum and so cardiac output is mainly dependent on heart rate. Acute hypoxia produces an immediate, reflex-mediated fall in FHR secondary to chemoreceptor stimulation of the vagus nerve (parasympathetic nervous system). However, if hypoxia is prolonged then adrenal stimulation produces an elevation of catecholamines that can overcome the vagal drive. If the hypoxia is corrected then the reflex parasympathetic drive reduces immediately leaving the elevated catecholamine level to produce a tachycardia, which settles over the subsequent 30–60 min. Thus, a moderate increase in FHR during labour, in the absence of a rise in maternal temperature, may be an indication of the sympathetic response to prolonged hypoxia and should be considered a warning sign.

The fetal response to a reduction in oxygen supply depends upon factors such as speed of onset, degree, and duration. When acute and severe, such as with cord prolapse or placental abruption, the FHR response is a prolonged bradycardia or recurrent, prolonged, decelerations. If fetal hypoxia evolves slowly, the fetal response will be modified by compensatory mechanisms, which depend upon the reserve capacity of the fetoplacental unit (intervillous space). Initially the FHR may not change but variability becomes reduced and accelerations disappear. Braxton Hicks tightenings or labour contractions may provoke late decelerations in such circumstances.

Fetal acidaemia

Besides redistributing blood to vital organs, the fetus can switch from aerobic to anaerobic metabolism in situations of hypoxia. Glucose is broken down to lactic acid, which, in the presence of oxygen, is converted to carbon dioxide (CO_2), releasing energy. CO_2 is transported to the placenta and diffuses into the maternal circulation for elimination. When this transfer is interrupted, such as by cord compression, hyprtonic uterine activity, or reduced uterine perfusion, CO_2 accumulates and forms excess carbonic acid. The subsequent formation of hydrogen ions (due to a shift of the equilibrium towards dissociation of H^+ and HCO_3^{2-}) results in fetal acidosis. This respiratory acidosis usually accompanies the hypoxia that results from any mechanism that interferes with fetal–maternal gas exchange, and constitutes the physiological definition of asphyxia. Initially, base deficit remains normal or is slightly increased; this acidosis can be readily reversed if CO_2 is eliminated.

With more severe hypoxia, respiratory acidosis progresses to a mixed respiratory and metabolic acidosis, with the latter increasing in significance the longer the hypoxia continues. As lactic acid accumulates so the pH will begin to fall further and the base deficit will rise. Inefficient glucose metabolism results in a rapid depletion of glycogen stores that, if normal, will allow the fetus to tolerate a short period of severe hypoxia (20–30 min), without developing severe acidosis.

A continuation, or worsening, of the situation leads to an acidosis dominated by the metabolic component, with further fall in pH and increase of base deficit. In contrast to problems in transporting CO_2 across the placenta, metabolic acidosis cannot be reversed unless oxygenation of the fetus can be re-established. The fetal tissues will accumulate acid metabolites, which will not be removed until the fetal circulation is re-established upon correction of the hypoxia. Fetal brain injury is likely if the hypoxia is severe or prolonged.

From a clinical point of view, it is important to try to estimate how far the hypoxic process has developed because, once the course of events has started, the fetal condition can deteriorate rapidly. Although the fetus can cope with respiratory or mild mixed acidosis for some hours, an acute fall in oxygen supply will produce a rapid shift to metabolic acidaemia and would not be tolerated for long.

Clinical interpretation of fetal scalp pH values

Neonatal condition at birth after intrauterine hypoxia is dependent upon the type of acidosis and stage of the hypoxic process. The FHR pattern may give an indication of the fetal condition but cannot be relied upon to predict the degree of fetal acidosis. This can be illustrated by a few examples. Cord compression may produce an FHR pattern with a fairly dramatic appearance of large variable decelerations. If prolonged, CO_2 accumulation may occur and a respiratory acidosis will develop. The baseline FHR usually remains normal and baseline variability indicates that the fetus is not significantly hypoxic. However, an unexperienced interpretation of severe decelerations, together with a preacidotic scalp blood pH value between 7.2 and 7.25, might provoke delivery. Such a baby is likely to have a normal Apgar score. Assessment of the complete acid–base balance will reveal the high CO_2 value and a normal base deficit, indicating respiratory acidosis. In another situation, the fetus may be chronically hypoxic with an FHR pattern of late decelerations (may be of mild to moderate amplitude), an abnormal baseline rate (in the range of 150–170 bpm) and reduced or absent variability. This is commonly associated with chronic placental dysfunction and fetal growth retardation. The acidaemia may not be very severe, with a pH between 7.15 and 7.2, but of a metabolic type with an increased base deficit. At birth the baby may be clinically depressed with low Apgar scores and be at risk of neurodevelopmental problems. The same pH value (in the range of 7.15 to 7.2) may also be associated with quite different clinical features and outcome in the newborn, depending upon the cause of the hypoxia.

Role of fetal scalp pH sampling

First, the complete acid–base balance should, if possible, be assessed to differentiate between respiratory and metabolic acidosis. Second, it should be recognized that the FHR, baseline variability, and fetal reactivity may give a better indication of fetal condition than the type or depth of decelerations. Third, a single fetal scalp pH assessment will give very limited information regarding the significance of a FHR trace abnormality. Significant hypoxia will produce a progressive acidaemia that will only be recognized by repeated measurements at short intervals. The FHR pattern during the intervals is of great importance. Baseline variability is crucial because absent short-term FHR variability (flat baseline) is indicative of fetal hypoxia (see Figs 15.1, 15.4 and 15.7).

It may not be appropriate to wait for the pH to reach an abnormally low level. A clear trend downwards, likely to take the scalp pH much below values between 7.15 and 7.2 before delivery, is an indication to deliver. It should be remembered that the suggested ranges of fetal scalp pH as preacidotic (7.2–7.25) and acidotic (<7.2) are the result of early studies on complicated, high-risk, pregnancies. Recent data indicates that significant neonatal morbidity is unlikely, in the absence of clinical risk factors, unless the umbilical artery pH is below 7.05.

Fetal blood pH is not the 'gold standard' of fetal condition but is only one parameter by which fetal tolerance to labour is assessed. As previously mentioned, a transient fall in pH of respiratory type is not uncommon and may lead to hasty obstetric intervention. In such situations, the fetus is usually still reactive to external stimuli, such as vibroacoustic stimulation.

The machinery error of measurement has a range of ±0.04. A fetal blood pH value of 7.22 could be anywhere between 7.18 and 7.26. Furthermore, if the fall in pH is transient, such as after a short bradycardia or oxytocin overstimulation, the fetus will compensate rapidly and may demonstrate a normal pH 30 min later (see Fig. 11.9). On the other hand, the pH of a hypoxic fetus may, during the same period of time, drop to a very low value. The FHR patterns are usually severely abnormal in such cases: loss of short-term variability being the most prominent feature, together with late or pronounced variable decelerations.

It is necessary to appraise a pH value (particularly a single one) with caution. A low pH value is unlikely if the FHR tracing is reassuring. In such circumstances a repeat sample should be obtained. FHR changes induced by oxytocin overstimulation would not usually call for fetal blood sampling if cessation of the oxytocin corrects the FHR. Fetal blood sampling should not be carried out during a prolonged deceleration. Delay in intervention may be contraindicated when trying to obtain blood from the fetus when the clinical situation (e.g. placental abruption) or the FHR pattern calls for immediate action (see Chapter 11). The same applies for the second stage, when an instrumental delivery may be more appropriate.

Normal range

The fetal scalp blood pH is normally 0.1 unit below the maternal blood value; pH, PCO_2, PO_2, HCO_3, and base deficit are stable during the first stage of labour. The mean pH values vary in the scientific literature between 7.3 and 7.4 during the early first stage of labour and between 7.2 and 7.37 during the late first stage of labour. However, the differences in mean values may be explained by the different populations studied, as well as by technical differences. We determined acid-base values in fetal scalp blood in 120 low-risk women in the first stage of labour at a cervical dilatation of 5 and 10 cm, and in cord artery blood at delivery before the first breath (Table 3.1). All samples were obtained by two investigators with the patient in a lateral position. None of the women had an epidural block for pain relief. Mean values were virtually unchanged (7.33–7.34) during the first stage of labour. The mean pH value in umbilical arterial blood was a little lower (7.29).

The distribution of different pH values in the population was of interest (Table 3.2). About 5 per cent of these low-risk cases had a scalp blood pH <7.25 either at 5 or 10 cm cervical dilatation. When acidosis (or preacidosis) was present it was usually of respiratory type and had recovered within 15–30 min. Transient low pH values in normal labour may result in unnecessary intervention. Out of nine cases with transient low pH values, a preacidosis of metabolic type was found in one case at 10 cm (pH 7.21, PCO_2 47 mmHg, HCO_3 18.1 mmol/l). The probable cause of the low pH was sampling when the head was crowning; a metabolic preacidosis is not uncommon at this stage of labour. In another case, maternal acidosis (pH 7.31) was the probable cause of fetal preacidosis

Table 3.1 Acid-base balance in scalp blood and cord arterial blood at birth

	Cervical dilatation (cm)					
	5		10		Cord artery	
	Mean	SD	Mean	SD	Mean	SD
pH	7.33	0.05	7.34	0.07	7.29	0.07
PCO_2 (mmHg)	46.85	8.15	46.57	7.36	50.56	9.47
PO_2 (mmHg)	25.00	12.69	24.06	13.58	21.05	8.71
HCO_3 (mmol/l)	24.15	2.67	23.93	2.88	23.08	2.69

Table 3.2 Distribution of pH values in scalp and cord arterial blood at birth

	Cervical dilatation (cm)				Cord artery	
	5		10			
pH	n	%	n	%	n	%
<7.100	0	–	0	–	2	1.7
7.100–7.149	1	0.8	2	1.7	2	1.7
7.150–7.199	1	0.8	1	0.9	5	4.2
7.200–7.249	5	4.2	4	3.4	22	18.3
7.250–7.299	24	20.0	21	17.8	30	25.0
7.300–7.349	42	35.0	48	40.6	35	29.1
7.350–7.399	41	34.2	32	27.1	18	15.0
7.400–7.449	6	5.0	8	6.8	6	5.0
>7.450	0	–	2	1.7	0	–
Total	120	100.0	118*	100.0	120	100.0

*Two patients had Caesarean section for fetal distress before reaching a cervical dilatation of 10 cm.

(pH 7.24). A false high fetal blood pH may occur with maternal hyperventilation as a respiratory alkalosis with blood pH values of 7.5 or more. However, some believe that pronounced maternal hyperventilation may have a deleterious effect on the uteroplacental circulation due to vasoconstriction or shunting in the placental bed and result in fetal hypoxia. The remaining seven cases had transient fetal blood preacidosis or acidosis of respiratory type with comparatively high P_{CO_2} values and normal HCO_3 values. In all cases the results within 30 min were normal—all cases had a reassuring FHR pattern. This illustrates the limited value of single pH samples. The pH needs to be interpreted in the context of the clinical situation. Maternal blood pH may be assessed in cases of low fetal blood pH in order to exclude maternal abnormalities.

Technique

Fetal scalp blood can be obtained once the membranes are ruptured and the cervix is dilated at least 2–3 cm. It is easier to collect blood if the cervix is effaced and the fetal head is engaged. The sample should be taken with the patient in lateral position, if possible, to avoid the effects of vena cava compression. An amnioscope of appropriate size for the cervical dilatation is inserted and the fetal scalp visualized and cleaned. Two incisions in the shape of a V are made and the blood collected in a capillary tube containing heparin. False low or high values can be obtained if the fetal blood is mixed with amniotic liquor or maternal blood. The collection of blood should be performed quickly to avoid the effects of a contraction. If the capillary tube is slow to fill (as may happen when filling the second half of the tube) it might be better to take a second tube and half fill it to the same level and then mix the two samples.

The sample should be analysed as soon as possible. It is not necessary to seal the capillary tube if the analysis is performed within a few minutes. If analysis is delayed, the ends of the capillary should be sealed after an iron filing has been inserted. Using a magnet to move the iron filing, the blood can be properly mixed and coagulation avoided. A cotton swab is pressed to the incision in the fetal scalp (with the amnioscope in place) for a few minutes to ensure haemostasis.

Umbilical cord acid-base measurement

Routine assessment of acid-base balance in umbilical cord artery (or cord vein) blood at delivery (before the first breath) is of great value for assessing the well-being of the newborn. Blood can be collected after the cord has been clamped. An easier way is to puncture the vessel when the cord is exposed at delivery with a fine needle connected to a 2 ml syringe containing heparin. The cord artery blood pH is usually 0.05 units below the vein blood value, although the difference might be greater in cases with breech presentation or after cord compression in labour.

Normal umbilical cord values of pH and blood gases have been reported from uncomplicated term pregnancies ending with normal FHR records during labour followed by vaginal delivery. Mean (SD) arterial values were pH 7.28 (0.05), P_{CO_2} 49.2 (8.4) mmHg, P_{O_2} 18.0 (6.2) mmHg and HCO_3 22.3 (2.5) mmol/l. Mean (SD) venous values were pH 7.35 (0.05), P_{CO_2} 38.2 (5.6) mmHg, P_{O_2} 29.2 (5.9) mmHg and HCO_3 20.4 (4.1) mmol/l.

Mean (SD) values of pH for total populations have been reported by a number of workers. There is broad agreement that the value is between 7.2 and 7.23 (0.07–0.08). Recent studies have shown that severe acidaemia at birth may not be significant unless the pH is less than 7.05, or even 7. The important outcomes by which these studies have evaluated the significance of pH at birth are neonatal neurological status and subsequent neurodevelopmental development. Umbilical acid-base state is not an appropriate end-point in itself and is poorly related to neonatal and childhood outcome unless very low (see Chapter 18). However, it provides an accurate measure of the effect of labour on fetal acid-base and blood gas values thereby giving an idea of the respiratory function of the placenta during labour.

4
Uterine activity

Introduction

Uterine smooth muscle is arranged in an outer longitudinal layer and an inner circular layer. This arrangement facilitates constriction of the penetrating uteroplacental arteries for haemostasis after delivery. However, during pregnancy and labour, constriction of these arteries during uterine contractions reduces maternal perfusion of the intervillous space.

Effect of contractions on intervillous perfusion

The intervillous space contains about 250 ml blood and, if the uteroplacental circulation is impaired, a reduction in oxygen tension in the intervillous blood may result in fetal hypoxia during each contraction. Each contraction may be regarded as a potential threat to placental circulation; the amplitude, shape, and duration will affect the degree of reduction in uteroplacental circulation (Fig. 4.1). The fetus is not stressed by normal contractions if placental circulation is normal (Fig. 4.2). However, abnormally strong or prolonged labour, especially a prolonged increase in basal tone, may reduce fetal oxygenation and cause fetal distress, despite a normal intervillous circulation. When fetal oxygenation is already impaired, as with placental insufficiency or in partial abruptio placentae, even normal labour may result in fetal hypoxia. In such circumstances, late

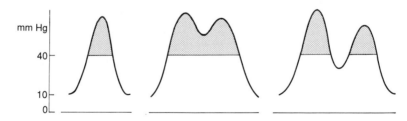

Fig. 4.1 Uterine contraction shapes. The hatched areas indicate the duration of reduced placental circulation when the amplitude of contractions exceeds 30–55 mmHg.

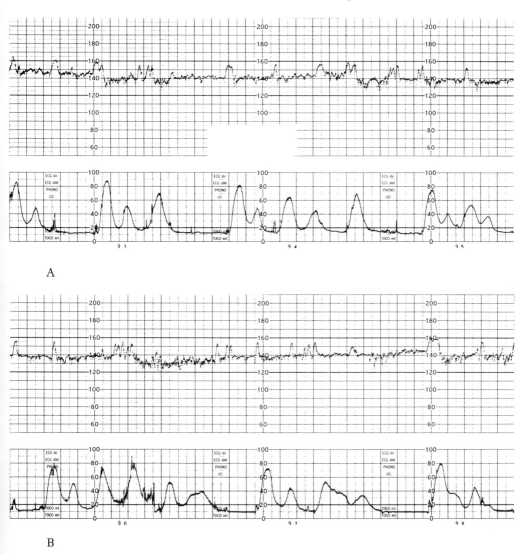

Fig. 4.2 An intrauterine pressure record of spontaneous uterine activity. The contraction pattern is abnormal but there is no effect on the FHR. Note that the amplitude of contraction is less when the basal tone at the start of a contraction is higher. (Panel B follows directly on from A.)

decelerations are often the response and this is the basis of the oxytocin challenge test (OCT). The OCT is not popular in the UK, where the preference, in such complicated pregnancies, is to rely upon the FHR response to spontaneous uterine activity (Braxton Hicks contractions).

Physiology of uterine activity

Uterine activity during pregnancy

Basic information in this field was obtained by Caldeyro-Barcia and co-workers in Montevideo. During the first 30 weeks of pregnancy two different types of contractions occur:

(i) short and frequent contractions of low intensity (up to 1 per minute) are triggered locally in the myometrium and are not perceived by the mother;
(ii) the other type of contraction (Braxton Hicks) involves the whole of the uterus, is of higher intensity (up to 10–15 mmHg), and is sometimes perceived by the mother.

Braxton Hicks contractions usually begin sporadically but, during the last 10 weeks of pregnancy, may occur more frequently, often creating anxiety in the mother-to-be. External monitoring of uterine activity may reveal regular contractions, which are palpable but are not perceived by the mother. Such monitored, painless, uterine activity seldom proceeds to preterm delivery (Fig. 4.3). As the onset of labour approaches they reach an intensity of 30–40 mmHg and become more frequent and regular (at least one every 10 min). However, factors governing the transition from Braxton Hicks contractions to the uterine contractions of labour are poorly understood.

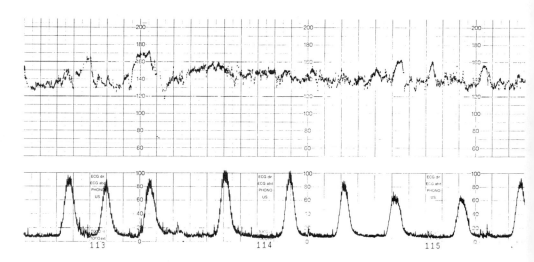

Fig. 4.3 An external record of preterm uterine activity at 35 weeks gestation. The tocograph suggests large and strong contractions but the mother had no sensation of pain. The contractions disappeared spontaneously and the pregnancy continued until term.

Labour contractions

The Montevideo group found that intrauterine pressure varied with the intensity of uterine activity at different levels of the uterus and with the degree of synchronization of uterine muscle. Normal uterine contractions start at the fundus and spread downwards to the lower uterine segment at a rate of 2 cm/s. The whole uterus becomes involved within about 15 s. The duration is longer at the fundus than at the lower uterine segment. Between contractions the basal tone in the uterine muscle is low.

The initiation of contractions is poorly understood. The Montevideo group postulated the presence of a pacemaker in each uterine cornu; this concept is now questioned. The lower uterine segment plays a passive role and does not contribute to the forces pushing the fetus into the birth canal. Failure to progress will result when contractions start in the lower segment and spread upwards and/or when the intensity is greater in the lower segment. Such complications cannot be diagnosed by a single intrauterine pressure catheter, as used in clinical practice.

Characteristics of contractions

The uterine activity record represents amplitude and duration, shape, frequency, and basal tone between contractions (Fig. 4.4).

Amplitude

The peak amplitude of contractions increases from between 30 and 50 mmHg early in the first stage of labour to 60 mmHg or more later in the first stage. In augmented labour an amplitude of 50 to 60 mmHg is usually effective. During the second stage of labour the amplitude rises even further and, together with maternal pushing, may even reach 150 mmHg or more. There is no recommended upper limit for the amplitude of contraction pressures but if the amplitude is greater than 70 mmHg then careful observations of maternal and fetal condition are advised.

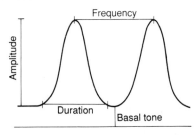

Fig. 4.4 An illustration of the different parameters used to quantify the uterine contraction record.

Duration

The duration of a contraction is an important parameter when assessing the efficacy of contractions. In the first stage of labour duration varies between 30 and 60 s. Later on, and in the second stage, it often lasts 90 s.

Shape

The shape of contractions is important. Ideally, the contraction curve should start from a low basal tone, rise steadily to a peak of adequate amplitude, slow down with a rounded appearance at the peak and then fall steadily back to a normal baseline tone. However, this is often not the dominant shape of contractions seen in spontaneous, normally progressing, labour; other shapes are often seen and should not lead to intervention unless there is failure to progress or fetal distress.

Two different types of contraction may dominate at different times during labour (Fig. 4.5). Type A predominates early in the first stage and is distinguished by a slow increase in amplitude with a swift return to basal tone. Type B, which is more common later in the first stage, increases rapidly in intensity at the beginning of the contraction and returns more slowly. This type is considered to be more efficient than the slow onset type. Despite the slow return, the duration is comparatively short because of the rapid increase at the start.

Contractions with double or triple peaks are abnormal because they lead to prolonged compression of the uteroplacental arteries and may predispose to fetal distress (Fig. 4.6). The fetal response is often recorded as a bradycardia or late decelerations. The subsequent peaks of such a contraction are often of lower intensity because the rise starts from a higher basal tone.

Frequency

Frequency is optimal when there are between 2 and 3 contractions per 10 min early in the first stage increasing to no more than 4 to 5 per 10 min later in the first stage of labour. A frequency of more than 5 contractions per 10 min is regarded as abnormal.

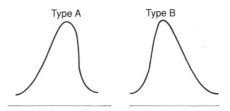

Fig. 4.5 An illustration of the two common types of contraction shapes.

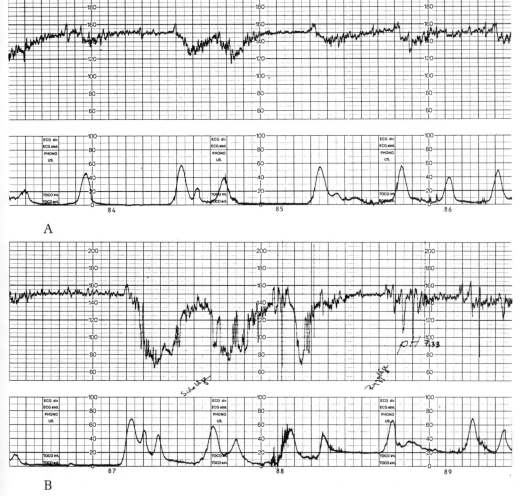

Fig. 4.6 An intrauterine pressure record of uterine activity. A, Normal contraction pattern for the first 30 min; B, triple and duplex contractions giving rise to variable decelerations. A fetal scalp blood sample was performed at mark 89 (pH 7.33). Note the apparent increase in basal tone when the mother lay supine for the scalp blood sampling.

Basal tone

Basal tone is approximately 10 mmHg early in the first stage, a range of 8–12 mmHg being acceptable. Basal tone rises later during the first stage reaching between 15 and 20 mmHg. An intrauterine pressure above 20 mmHg between contractions is sufficiently high to cause concern, even though this may be recorded in apparently normal labours without fetal distress.

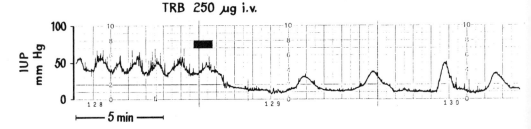

Fig. 4.7 An intrauterine pressure record of uterine activity showing severe hyperstimulation (baseline pressure between 30 and 40 mmHg) secondary to oxytocin. At TRB a bolus injection of terbutaline 0.25 mg i.v. was given to relax the uterus. Normal basal tone and normal contractions followed.

Hyperstimulation

If the contraction rate is too frequent then uterine activity may be seen as a continuous wave oscillating from an elevated baseline. This is often iatrogenic (induction or augmentation) and is most threatening to the fetus. Such pathological uterine activity may be abolished by the administration of a tocolytic drug (intravenous infusion or bolus injection of a β-receptor agonist, such as terbutaline, salbutamol, or ritodrine) (Fig. 4.7).

Prolonged contractions may also occur spontaneously. Pregnant women sometimes report such contractions during pregnancy and, if they occur during monitoring, the fetus may well be seen to respond with a prolonged deceleration. Before and after such a contraction the record is usually quite normal. This

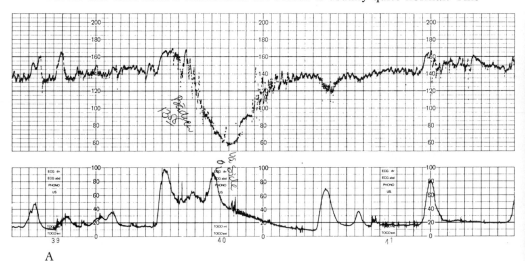

Fig. 4.8 (caption opposite).

Physiology of uterine activity 41

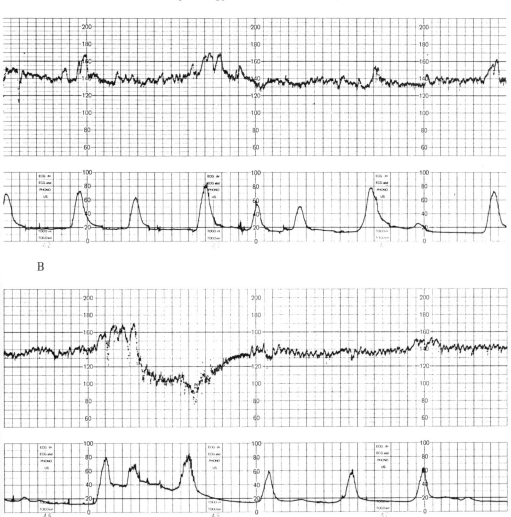

Fig. 4.8 Intrauterine pressure and direct (fetal scalp) FHR. A and B, an episode of abnormal uterine activity, whilst on the bed pan, resulted in a prolonged deceleration followed by a normal FHR and contractions; C, an episode of spontaneous abnormal uterine activity, with basal tone exceeding 30 mmHg, produced another prolonged deceleration.

is an occasional, isolated, phenomenon and the fetal response should be regarded as a normal reaction and not distress (see Figs 14.24 and 14.25). Similarly, in normal labour, spontaneous abnormal contractions occasionally give rise to a prolonged deceleration (Fig. 4.8).

Monitoring uterine activity

Palpation

External palpation is the oldest method of estimating uterine activity. The frequency may be ascertained quite accurately but duration and intensity are more difficult to assess. The uterus cannot be imprinted by the fingers of the examiner's hand during a good contraction. The mother's sensation of pain is a good guide to frequency but is not an accurate guide to intensity. Mechanically, a small uterus will produce a higher intra-amniotic pressure than a larger one, according to the law of Laplace, which states that pressure is inversely proportional to the radius of a cavity. Measurements from the non-pregnant uterus have shown values as high as 200 mmHg without any sensation of pain. In patients at term, intra-amniotic pressures of between 70 and 80 mmHg are occasionally registered without maternal discomfort.

Manual palpation and external tocography do not give any guide to intensity and are, therefore, likely to be misleading and result in unnecessary augmentation of labour (Fig. 4.9). If provoked, decelerations (often combined) are often an indication of uterine overstimulation. Intrauterine pressure monitoring is appropriate when augmentation of labour is difficult.

The position of the patient has an important influence on the intensity of contractions. In the supine position, contractions are more frequent but of lower

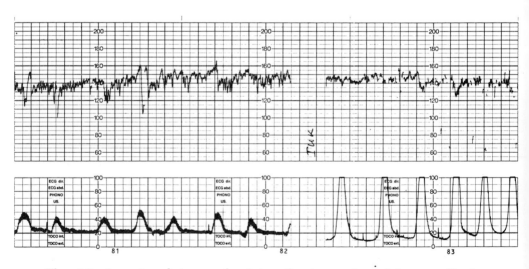

Fig. 4.9 An external tocograph giving the impression of low amplitude contractions followed by evidence of severe hyperstimulation. Contraction amplitudes exceeding 100 mmHg were seen after insertion of an intrauterine pressure catheter.

amplitudes. In the lateral and erect positions, contractions are less frequent but stronger. Parturiants should be encouraged to avoid the supine position.

External monitoring of uterine activity

A tocodynamometer, fastened by an elasticated band around the maternal abdomen at fundal height, is the usual method of continuous monitoring. Being non-invasive, the method can also be used during pregnancy. Monitoring the onset of a contraction is useful for mothers who wish to prepare before the onset of the pain.

Duration, as well as frequency, may be determined with reasonable accuracy. Intensity cannot be ascertained by external methods. Accordingly, no reference to mmHg is possible with external monitoring. The external record is dependent upon the thickness of the abdominal wall, the position of the mother, and the tension of the fastening band. The tocodynamometer sometimes has to be relocated during labour to find the optimal site for registration.

Internal monitoring of uterine activity

Several methods have been used to measure pressure within the uterine cavity during labour. Most commonly, a fluid-filled catheter within the amniotic cavity is connected to a pressure transducer. Alternatively, there is a catheter with a pressure transducer at its tip.

Sets for internal monitoring are commercially available. Membranes need to be ruptured and the cervix 1–2 cm dilated. The catheter should pass easily between the cervix and the presenting part of the fetus, although a deeply engaged fetal head may create difficulty. If there is resistance, a new location should be tried. At insertion, slight bleeding may occur but serious complications, such as haemorrhage due to rupture of placental or fetal vessels, or uterine perforation, are extremely rare and unlikely if force is avoided.

With the catheter in position the transducer should be fixed at the level of the symphysis pubis. The transducer needs to be set at zero in atmospheric pressure and calibrated against a known pressure. Technical problems include air bubbles in the transducer, leakage of fluid from the system, or catheter block by mucus or detritus.

It is important to recognize technical artefacts on the record, such as sudden, sharp, changes up or down or a totally straight baseline indicative of a partial or complete block of the catheter (Fig. 4.10). Other deviations on the toco trace result from other mechanisms for increasing intrauterine pressure, such as coughing, heavy breathing, or pushing.

The catheter-tip transducer is also commercially available as a disposable catheter that is easy to calibrate and gives accurate recordings without artefacts related to catheter blockage (Fig. 4.11).

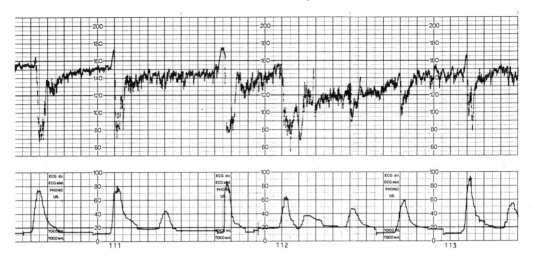

Fig. 4.10 The toco record shows contractions of irregular shape due to partial blockage of the intrauterine pressure catheter.

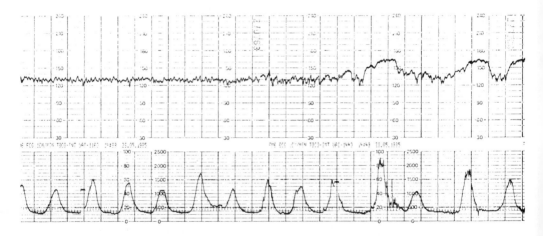

Fig. 4.11 Uterine activity recorded using a catheter-tip intrauterine transducer. A quantitative estimation of total intrauterine pressure (UAI) is indicated by a short horizontal bar on the toco record every 15 min by calculation of the area under the contraction trace (Uterine Activity Integral).

Indications for intrauterine monitoring

Internal monitoring should be considered when progress in labour is abnormal. Intrauterine pressure measurement will be useful, in such cases, to help differentiate between dystocia secondary to poor uterine activity and true

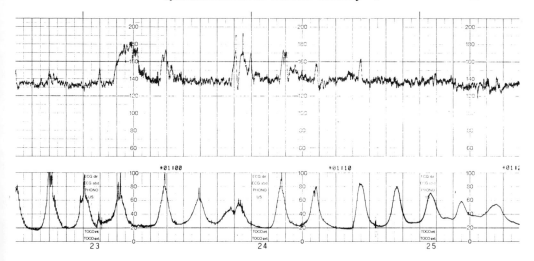

Fig. 4.12 An intrauterine pressure record of spontaneous uterine activity showing basal tone just above 20 mmHg. Delivery was ultimately by Caesarean section because of dystocia.

obstruction despite good uterine activity. This is specifically important in labour with increased tone (hypertonic labour), although baseline tone may be difficult to register precisely as the position of the catheter tip within the uterus is unknown (Fig. 4.12).

Another indication for considering intrauterine pressure monitoring is when labour requires induction or augmentation. Of particular value is the use of real pressure measurements to titrate the uterine response to oxytocin. Different responses depend on individual sensitivity to oxytocin. In some circumstances, even a small increase in oxytocin (e.g. from 2 to 4 mU/min) may result in overstimulation leading to fetal heart rate changes such as late and combined decelerations (Fig. 4.13). Late in the first stage of labour, uterine sensitivity to oxytocin tends to increase as labour progresses. We found the majority of combined decelerations (often related to induced abnormal uterine activity) appeared during the late first stage when the cervix was dilated between 7 and 10 cm.

Finally, intrauterine pressure monitoring should be considered for women in labour after a previous Caesarean section, especially after epidural block has been given or if labour requires augmentation (Fig. 4.14).

Quantification of uterine activity

There have been several attempts to quantify uterine activity in order to assess adequacy of labour. Montevideo units (MU) are most commonly used for

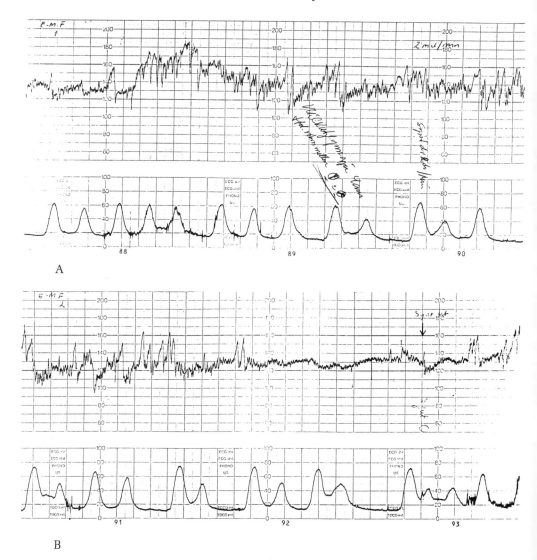

Fig. 4.13 (caption opposite).

assessment and are defined as the mean intensity of contractions multiplied by the frequency during a 10-min period (Fig. 4.15). During the latent phase it is not unusual to see uterine activity amounting to 50 MU before established labour; early in the first stage, 120 MU are often recorded; late in the first stage, figures around 200 MU are common; values above 250 MU are considered pathological. Montevideo units do not take duration into consideration.

Quantification of uterine activity 47

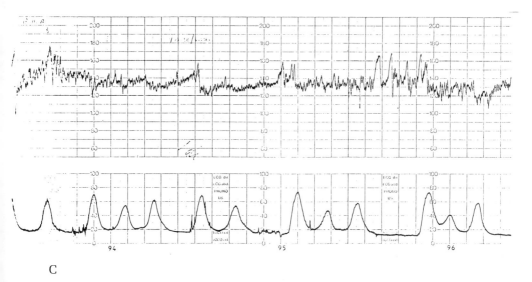

C

Fig. 4.13 An intrauterine record of uterine activity and direct FHR monitoring. A, An infusion of oxytocin (2 mU/min) was started at the end of panel (at mark 90); B, abnormal contraction pattern as a result of hyperstimulation until the infusion was stopped at mark 93; C, abnormal contractions seen again after recommencing the oxytocin infusion at 1 mU/min at mark 94.

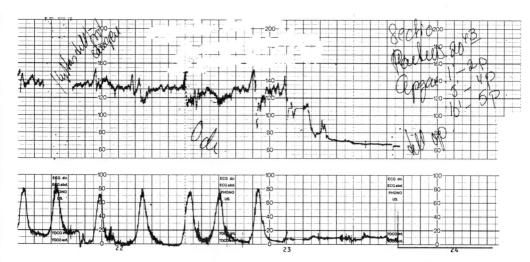

Fig. 4.14 An intrauterine pressure record showing loss of tone after scar rupture in a patient with a previous caesarean section. The spontaneous contractions suddenly ceased at mark 23 and, simultaneously, a severe FHR bradycardia, with absence of baseline variability, appeared. The scar rupture was verified at emergency Caesarean section. The baby was depressed at birth, with Apgar scores of 2, 4, and 5.

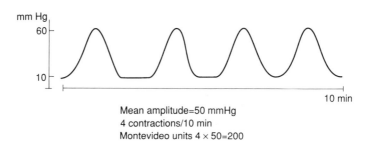

Fig. 4.15 Calculation of Montevideo units of uterine activity.

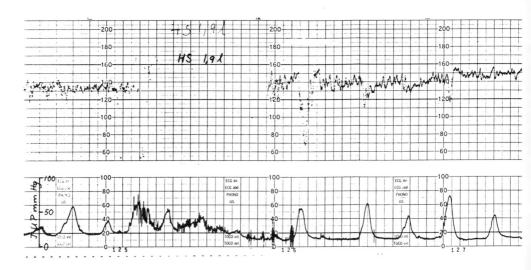

Fig. 4.16 A case of polyhydramnios. A percutaneous intrauterine catheter shows that basal tone decreased after amniotomy allowed drainage of 1.9 litre of amniotic fluid (written 'HS' after mark 125). It is notable that the amplitude of the contractions increased when they started from a lower level of basal tone.

Alexandria units do this by being a product of mean intensity multiplied by frequency multiplied by mean duration per 10-min period.

Recently, more sophisticated methods have become available for assessment of uterine activity. The area of each contraction can be calculated electronically (see Fig. 4.11) and studies using this technique have shown that greater pressures are generated during induced compared with spontaneous labour. Uterine activity in spontaneous labour shows considerable individual variations. Activity is more intense in multiparous women, mostly because of higher frequency of contractions. Uterine activity increases from early first stage to late first stage by about 50 per cent, both in primiparae and multiparae.

Abnormal uterine activity

Hypoactivity is the mildest form of abnormality with an amplitude of contractions less than 30 mmHg and a frequency of less than two per 10 min. Basal tone is low or normal. Total uterine activity is below 100 MU. This abnormality is often corrected by infusion of oxytocin. In cases of polyhydramnios or multiple births the distended uterine muscle often produces low intensity contractions. Amniotomy and reduction of intra-amniotic pressure may improve labour progress (Fig. 4.16).

Vaginal or intracervical application of prostaglandin for cervical ripening might induce excessive uterine activity (Fig. 4.17). Cases with placental insufficiency or with previous Caesarean section should be especially monitored for signs of hyperstimulation.

Abruptio placentae often precipitates abnormal uterine activity, with a steep rise in intensity and frequency of contractions. The basal tone may be abnormally high and show a wavy contraction curve with a high frequency of contractions (Figs 4.18 and 4.19). Hyperactivity may also appear spontaneously and internal monitoring will make the diagnosis. Hypertonic abnormal uterine activity is characterized by a basal tone greater than 20 mmHg, while intensity and frequency might be normal. This abnormality can be deleterious for the fetus if it impairs placental circulation. Too rapid an infusion of oxytocin is a common cause (Fig. 4.20).

Summary

In summary, there is wide range of individual variations regarding intensity, duration, and frequency of contractions, even in normal labour. Uncoordinated or atypical patterns of contractions are often monitored in labour but are without importance as long as the progress of labour is normal. This emphasizes the fact that internal pressure monitoring will give no further useful information in the presence of normal progress and a normal FHR. Labour with normal progress may be adequately monitored by manual palpation or by external tocography. The latter has some advantages to manual palpation, being more accurate and giving a continuous print-out. Time relations between FHR changes and decelerations are readily visualized.

50 Uterine activity

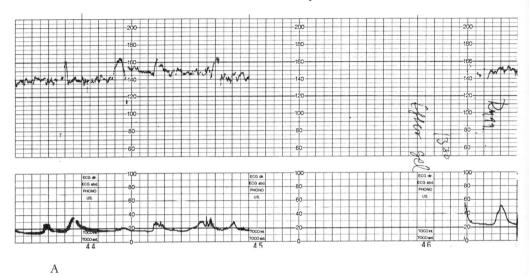

A

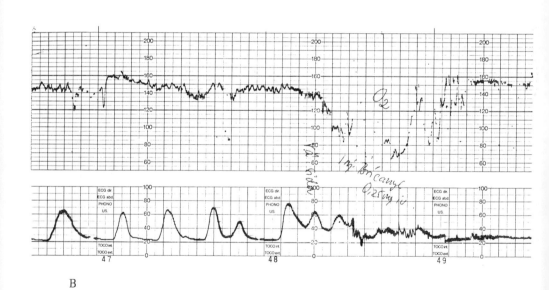

B

Fig. 4.17 (caption opposite).

Summary

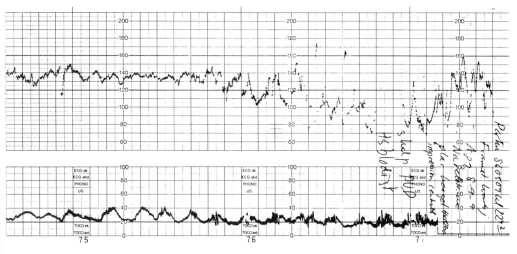

C

Fig. 4.17 An external record of uterine activity and FHR during induction of labour. A, The CTG prior to vaginal prostaglandin gel (at mark 46); B, hyperactivity of the uterus seen 20 min later causing a prolonged FHR deceleration. The uterus was relaxed with an intravenous injection of terbutaline (0.25 mg); C, in the second stage of labour, 5 h later, blood-stained amniotic fluid was seen following amniotomy. Spontaneous delivery occurred soon after and it was clearly evident that a small placental abruption had occurred. A blood clot measuring 4 × 10 cm was found on the maternal surface of the placenta. The Apgar scores were normal.

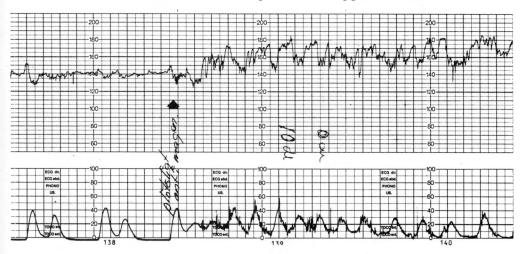

Fig. 4.18 A primigravida in spontaneous labour at term who reported a sudden onset of severe abdominal pain (indicated by the arrow). Uterine activity and FHR pattern changed simultaneously. At Caesarean section a placental abruption 3 × 3 cm in size was found.

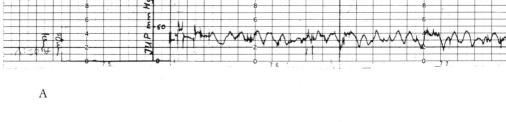

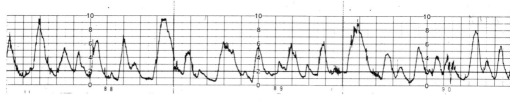

Fig. 4.19 An intrauterine record of uterine activity in a case with several placental abruption and a dead fetus. A, A high basal tone (20–30 mmHg) and an abnormal contraction pattern; B, after 1.5 h, the basal tone was 10–15 mmHg and contractions were less frequent but with higher amplitude.

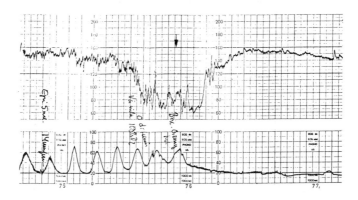

Fig. 4.20 An intrauterine record of uterine activity during augmentation of labour with oxytocin (10 mU/min). Hyperstimulation, with increased basal tone, resulted in a prolonged deceleration. The infusion of oxytocin was stopped immediately but without effect on uterine activity and FHR. A bolus of terbutaline 0.25 mg i.v. (indicated by the arrow) relaxed the uterus within 2 min and the FHR pattern returned to normal.

5
Control of fetal heart rate variability

Introduction

Knowledge of some basic facts about control of the FHR is important for an appropriate interpretation of the CTG trace. Influences such as the autonomic nervous system, hormonal factors, and metabolic condition, vary during the course of pregnancy as the heart develops and control of FHR matures. Baseline FHR is highest (175 bpm) in the ninth week of pregnancy, falls to about 160 bpm by 15 weeks and varies between 120 and 160 bpm from 20 weeks onwards.

Autonomic nervous system influences

The fetal heart is innervated by nerve fibres of the autonomic nervous system. Parasympathetic nerves are present by the eighth week of pregnancy, although the parasympathetic nervous system appears not to exert full influence until after 32 weeks. Increasing vagal tone towards term is a major influence on the baseline FHR. Stimulation of the parasympathetic nervous system produces a rapid, vagally mediated, fall in FHR. The rapidity of this fall contrasts with the result of sympathetic stimulation, which produces a slow rise in FHR. Stimulation of the sympathetic nervous system also improves myocardial contractility. Thus, the baseline FHR at any one time is under the influence of both parasympathetic and sympathetic activity.

Baroreceptors and chemoreceptors are important for the genesis of different types of decelerations. The FHR control is dominated by neural reflexes mediated via baroreceptors and chemoreceptors. These peripheral receptors are predominantly located in the arch of the aorta and the carotid sinus. The reflexes are modified in the medulla oblongata (brain stem), where central chemoreceptors are also believed to be located. Baroreceptors are stretch receptors responsive to changes in the blood pressure. An increase in blood pressure induces a reduction in the FHR mediated by increased vagal tone and inhibition of sympathetic activity. A reduction in blood pressure will increase the FHR by inhibition of vagal activity. Chemoreceptors respond to changes in the partial pressure of dissolved oxygen in fetal blood and the effect depends upon the difference between central and peripheral chemoreceptor stimulation. The carotid sinus (peripheral) receptors are activated by hypoxaemia before tissue hypoxia (central) is established in the brain and heart.

Other influences on FHR

Among other factors that affect FHR are the partial pressure of oxygen in fetal blood, haemodynamic factors (mainly blood pressure), fetal breathing, and the degree of fetal activity. Hormonal factors include the fetal adrenal glands, which secrete noradrenalin and adrenalin with isotropic and chronotropic effects on the heart muscle, corticosteroids, thyroid hormones, and insulin, which are important for growth and development.

FHR variability

Baseline FHR varies constantly and produces an irregular, saw-toothed, appearance on the CTG trace, which is described as baseline variability. Discrete deviations from the baseline rate, whether upwards (accelerations) or downwards (decelerations), are usually considered as separate events. Baseline variability has two major components, which are described as short-term and long-term variability. Short-term variability refers to differences between consecutive pulse intervals (pairs of heart beats) and can only be derived from the R–R intervals of fetal electrocardiogram (ECG) signal (Fig. 5.1). Long-term variability represents the fluctuations of the baseline rate over a longer time period, usually minutes, and may be described in terms of oscillation frequency (normally 2–6 per min) (Fig. 5.2) and amplitude (beats per minute) (Fig. 5.3).

Oscillation frequency is not practical for clinical use and so band-width, as an expression of the amplitude of the macrofluctuations, remains the standard method for assessing variability. Hammacher's original description of variability was:

(1) silent pattern <5 bpm (Fig. 5.4);
(2) reduced variability 5–10 bpm (Fig. 5.5);

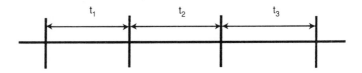

Fig. 5.1 An illustration of short-term FHR variability, by which is meant differences in consecutive pulse intervals (t_1, t_2, t_3, etc.).

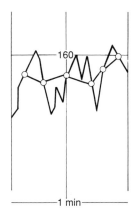

Fig. 5.2 Baseline FHR oscillations, the frequency of which have been used as a measure of FHR variability.

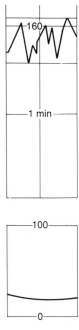

Fig. 5.3 Baseline FHR oscillations, the amplitude (bandwidth) of which is commonly used as a visual description of long-term FHR variability.

(3) normal 10–25 bpm (Fig. 5.6);
(4) saltatory >25 bpm (Fig. 5.7).

However, such descriptions are not commonly used now, and variability is referred to as absent, low, normal, or increased. Visual assessment of variability

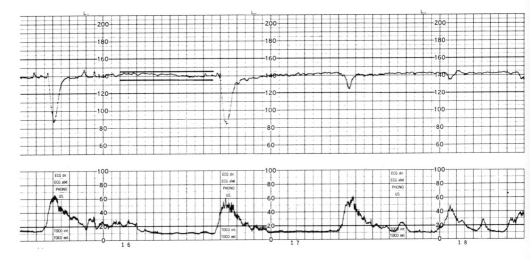

Fig. 5.4 Very low FHR variability (less than 5 bpm amplitude). The two horizontal lines at mark 16 indicate an amplitude of 10 bpm (125–135 bpm) for reference.

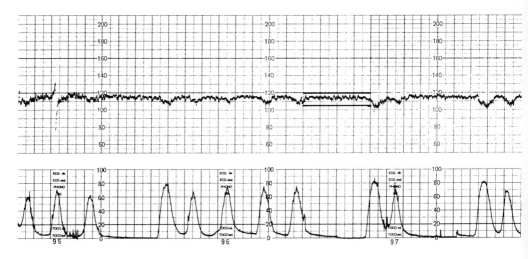

Fig. 5.5 Low FHR variability with an amplitude between 5 and 10 bpm. The two horizontal lines before mark 97 indicate an amplitude of 15 bpm (105–120 bpm) for reference.

has good interobserver agreement but correlation between visual and computerized assessment is poor, especially for short-term variability.

FHR variability is not entirely the product of neural control. After pharmacological blockade of the autonomic nervous system some FHR

FHR variability

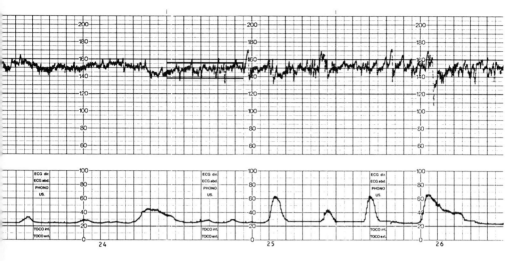

Fig. 5.6 Normal FHR variability between 10 and 25 bpm. The two horizontal lines before mark 25 indicate an amplitude of 20 bpm (between < 140 and < 160 bpm).

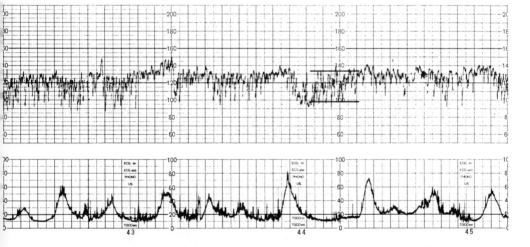

Fig. 5.7 High FHR variability, greater than 25 bpm. The two horizontal lines at mark 44 indicate a variability of 135 bpm (between 100 and 135 bpm).

variability can be seen. Hormones such as catecholamines also influence variability. With increasing gestational age and a more mature autonomic system, short-term and long-term variability increase, possibly as a result of increasing vagal dominance. When interpreting CTG traces in the early third trimester (28–32 weeks) this must be taken into account. Low variability is often seen in this period. The relationship between pulse intervals (in milliseconds) and FHR is shown in Figure 5.8.

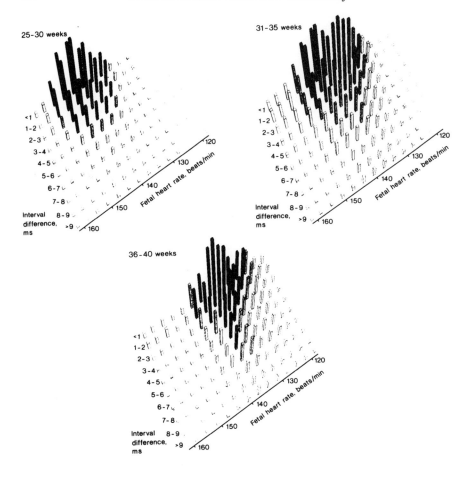

Fig. 5.8 Correlation between pulse interval differences and FHR at various gestations of pregnancy (from Westgren, personal communication).

Short-term and long-term variability

FHR variability is an important parameter of the CTG trace. However, there is much confusion in the literature regarding terminology and, consequently, interpretation of variability. Both short-term and long-term variability are recorded accurately if the monitor is connected to the fetus via a skin electrode that picks up the ECG. Short-term variability will then appear as very rapid changes (microfluctuations) of the baseline recorded on the trace (Fig. 5.9). The FHR itself determines the number of pulse interval differences and it is impossible to recognize visually so many microfluctuations of the baseline (equal in number to the FHR) when printed on paper running at 1 cm per minute.

FHR variability

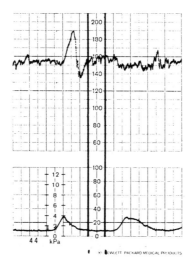

Fig. 5.9 Baseline FHR variability should be assessed between contractions to avoid their effect on the FHR.

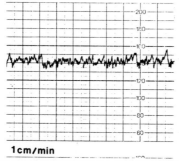

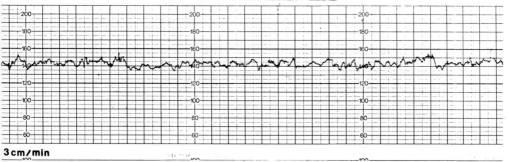

Fig. 5.10 A 10-min FHR record comparing the visual effect of two paper speeds — 1 cm per min and 3 cm per min.

Microfluctuations are seen more easily if the paper speed is faster (Fig. 5.10) but quantitative measurement of the short-term variability can only be obtained by computer analysis. Visual interpretation will, however, recognize a pronounced reduction in short-term variability and this may be associated with a concomitant reduction in long-term variability.

The relationship between short-term and long-term variability is complex. Normal variability indicates that the control mechanisms of the fetal heart (primarily neural reflexes) are intact. Long-term variability is generally considered to be the result of the interplay between parasympathetic and sympathetic activity whereas short-term variability is believed to be, principally, the result of fluctuations within the parasympathetic system.

There is a close correlation between short-term and long-term variability. Both are absent when the trace is flat and fixed, such as with congenital heart block (Figs 5.11 and 5.12). Reduced or absent variability indicates impairment of the normal neural or cardiac influences on FHR and can be seen in cases with profound hypoxia (Figs 5.13, 5.14 and see 5.18). For short periods the short-term variability may be low with normal long-term variability in the absence of fetal hypoxia (Fig. 5.15). Low short-term variability where long-term variability is seen as a smooth undulating wave form is usually recognized as a sinusoidal pattern.

Sinusoidal pattern

With this uncommon appearance the trace shows a regular, undulatory, waveform. In the literature there is no consensus as to how to define this pattern

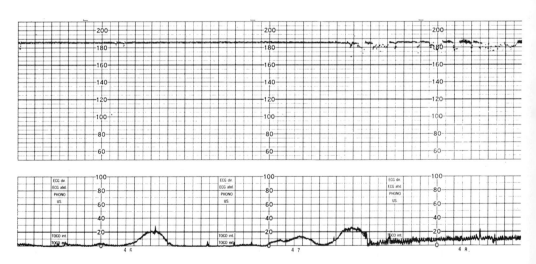

Fig. 5.11 The effect on FHR variability at the fast rate of atrial flutter.

FHR variability

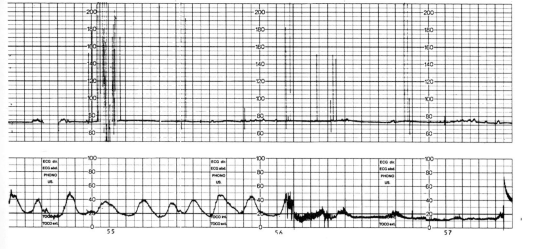

Fig. 5.12 Fetal atrioventricular heart block showing bradycardia with no variability.

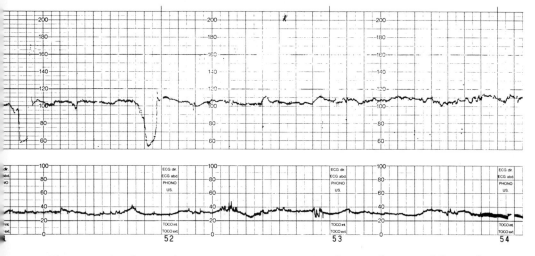

Fig. 5.13 An abnormal (ominous) antepartum CTG showing low variability and late decelerations associated with two Braxton Hicks uterine contractions (tightenings).

and published cases do not all fulfil the same criteria. As a consequence, published reports contain varying prognoses for the baby.

The following features have been described:

(1) regular oscillations above and below a normal baseline rate;
(2) amplitude of oscillations rarely greater than 5–15 bpm;

62 Control of fetal heart rate variability

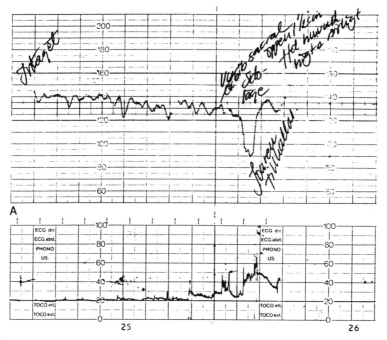

5.14A

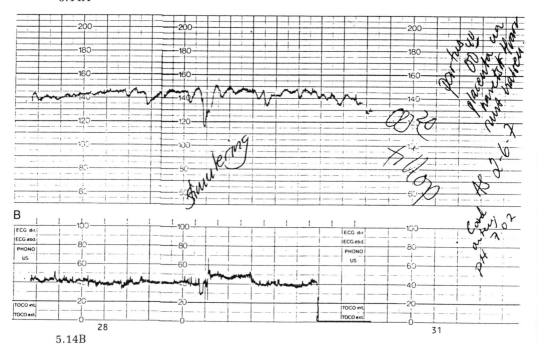

5.14B

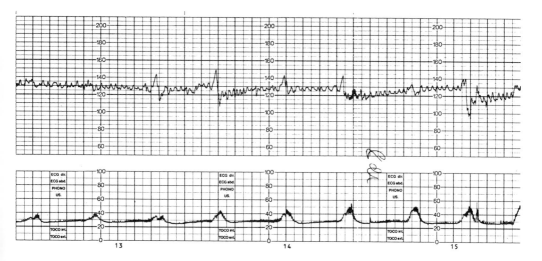

Fig. 5.15 Normal long-term variability but reduced short-term variability of doubtful significance (pseudosinusoidal).

(3) frequency of oscillations between 2 and 5 cycles per min;
(4) absent short-term variability;
(5) absent reactivity (Fig. 5.16).

The origin of this pattern is not fully understood. Normal neural control of the heart is probably absent. Tissue hypoxia in the fetal brain stem brain may be important for the genesis of this pattern. Fetal anaemia and severe asphyxia have been found associated with 'true' sinusoidal pattern. Rhesus isoimmunization (Fig. 5.17) is the most important cause of fetal anaemia but other causes include fetomaternal haemorrhage and fetal bleeding. In animal studies (fetal sheep experiments), where the fetus has been exposed to bleeding, a sinusoid pattern has been elicited. It has also been observed that a sinusoid pattern can disappear in rhesus cases after intrauterine blood transfusion.

Fig. 5.14 (opposite) A multipara at 40 weeks gestation who presented with reduced fetal movements. A, Admission test showing normal baseline FHR and no decelerations but absent short-term variability and normal long-term variability. The FHR response to vaginal examination was a deceleration; B, the CTG 10 min later showing persistence of the abnormal pattern. Vibroacoustic stimulation also produced a deceleration (at mark 28). An emergency Caesarean section resulted in a live birth with Apgar scores of 2, 6, and 7. The umbilical artery results were pH 7.02, P_{CO_2} 11.7 kPa, and base deficit 13.1 mmol/l, indicating a severe mixed respiratory and metabolic acidosis. The umbilical cord was four times around the neck. The baby was normal at 1 year of age.

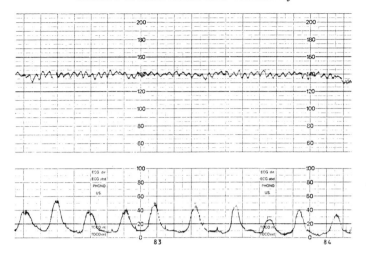

Fig. 5.16 A pattern often considered sinusoid.

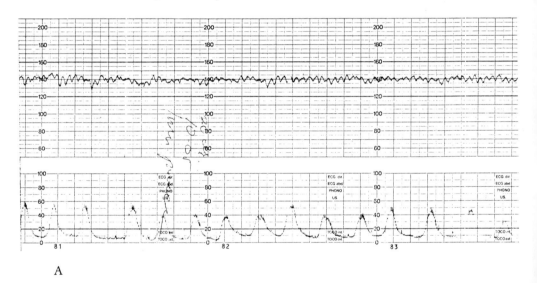

A

Fig. 5.17 (caption opposite).

Modanlou and Freeman (1982) reviewed the literature and found 27 published cases fulfilling the above-mentioned criteria. The perinatal prognosis was dismal; 24 cases were associated with antepartum or neonatal death or severe morbidity; 13 were cases of rhesus isoimmunization, three fetuses had severe anaemia with hydrops and seven had severe perinatal asphyxia. Of the three with a good outcome, alphaprodine had been administered to two mothers during labour and another fetus had gastroschisis.

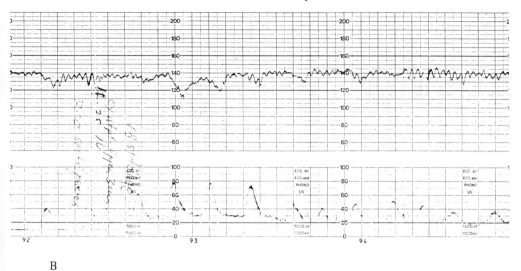

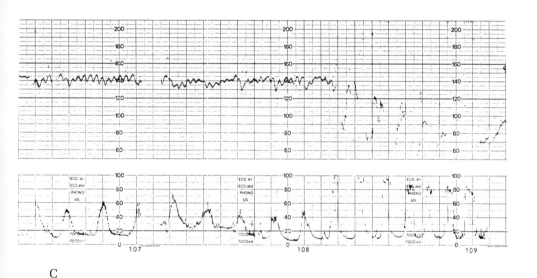

Fig. 5.17 A multipara aged 29 years with severe rhesus isoimmunization. Labour was induced at 37 weeks by amniotomy and oxytocin infusion. A, A sinusoid FHR pattern soon after amniotomy; B, after 90 min the same baseline pattern persisted but late decelerations were now also present; C, the CTG during the second stage just prior to shoulder dystocia at delivery. The Apgar scores were 4, 5, and 7. The neonate was severely anaemic.

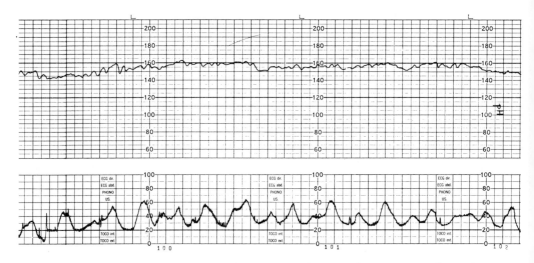

Fig. 5.18 A primipara at 32 weeks in preterm labour. The CTG trace shows the last 30 min of a 3-h sinusoid pattern. The fetal scalp pH was 7.1 and so an emergency Caesarean section was performed. The umbilical artery pH was 7.09 and the Apgar scores were 1, 2, and 3. The baby had multiple lethal congenital malformations.

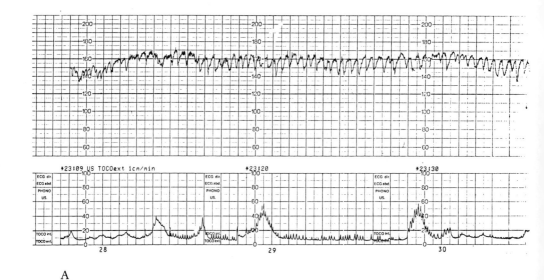

A

Fig. 5.19 (caption opposite).

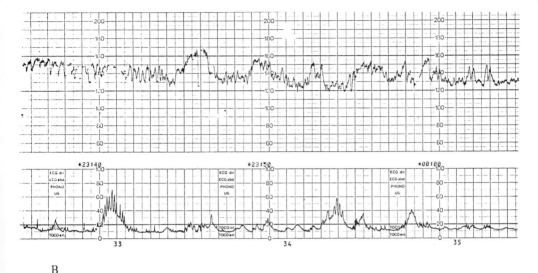

Fig. 5.19 An FHR pattern sometimes referred to as sinusoid because of the regularity of the FHR oscillations. Continuation of the record confirmed a subsequent normal FHR pattern.

Figure 5.18 shows a CTG record with a sinusoid pattern in a fetus with severe intrauterine acidosis (scalp blood pH 7.10). It is important not to include cases showing reactivity or normal variability in the FHR trace, such as seen in Figure 2.16. Figure 5.19 is another example of confusion with the sinusoid pattern. However, the sinusoid-like pattern was present for only a short period of time and the trace also shows normal reactivity and variability. Such traces have been related to rhythmic movements of the fetal mouth.

FHR reactivity and fetal behavioural states

The graphic record of FHR reflects many aspects of fetal life including body movements, breathing (chest) movements, and sucking (mouth) movements. FHR variability is influenced by these events, and accelerations and decelerations may also be related. Reactivity is a term used to describe FHR accelerations associated with fetal body movements and is a clear indication of fetal wellbeing. The behavioural cycle of newborn infants has served as a model for fetal behavioural studies and four levels of activity are recognized:

1. State 1F (quiet state): absence of fetal body movements perceived by the mother (quiescence) resembling the neonatal deep sleep state without rapid eye movements (REM).

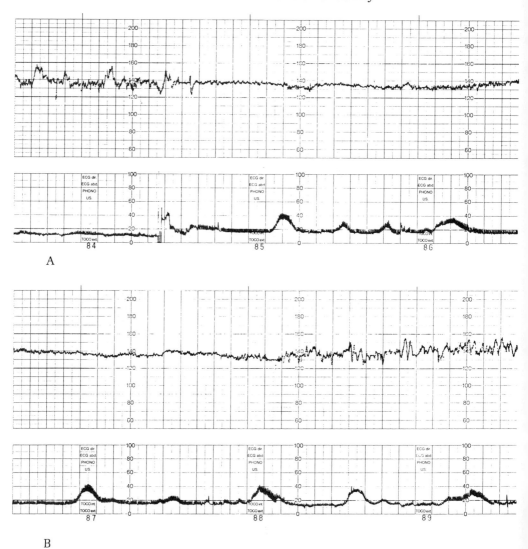

Fig. 5.20 An episode of low FHR variability indicative of fetal quiescence during pregnancy. Normal FHR variability and reactivity end after mark 84 in A and return at mark 88 in B.

2. State 2F (active state): fetal activity resembling the neonatal sleep state with REM.
3. State 3F (transition state): an intermediate state between fetal activity and true fetal wakefulness resembling the quiet awake state in the neonate.

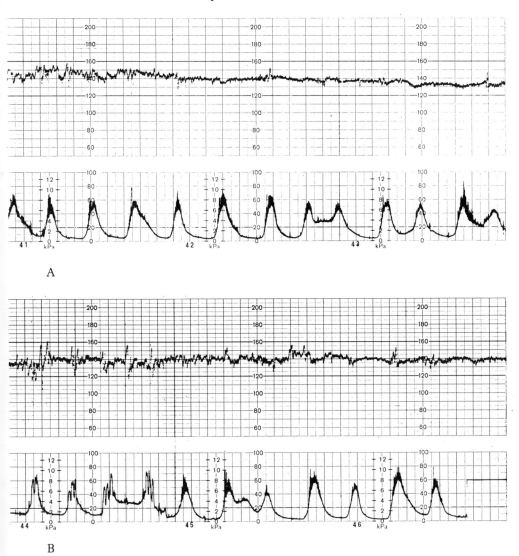

Fig. 5.21 Cyclic alternations between 'low' and 'high' FHR variability, indicative of quiet (after mark 42 in A) and active (after mark 44 in B) fetal behavioural states, during labour.

4. State 4F (awake state): true wakefulness resembling the active awake state in the neonate.

Behavioural states develop with gestational age and are not clearly discernible until 36 weeks. However, a more basic fetal rest–activity cycle, comprising

Control of fetal heart rate variability

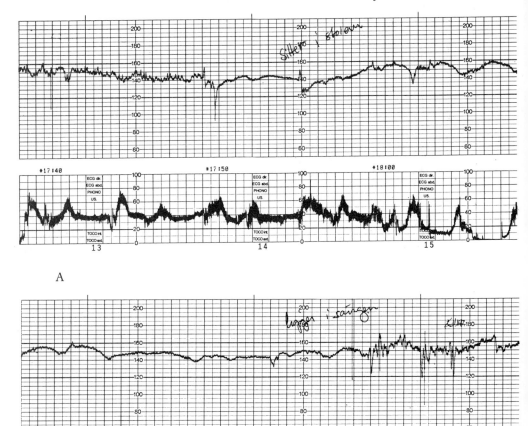

Fig. 5.22 Fetal quiescence with low variability and small decelerations for 40 min (1750–1830) during labour.

alternating episodes of quiescence and activity, is evident throughout most of the third trimester and lasts between 60 and 90 min in term pregnancies. The fetus spends about three-quarters of the time in state 2F (active). A diurnal rhythm of fetal movements also occurs with the highest frequency of movements in the evening and early night. During the day the frequency of fetal movements

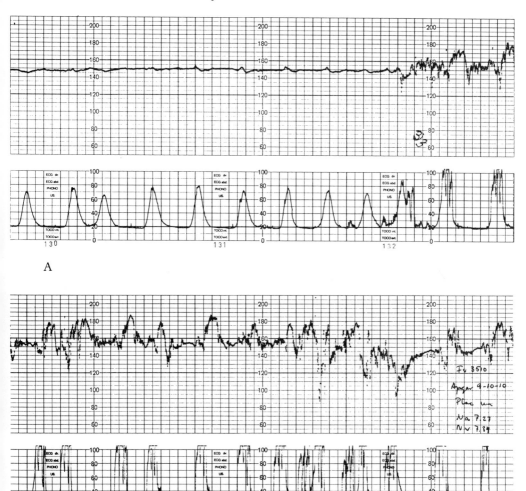

Fig. 5.23 A, Very low variability in the first part of this labour CTG record changed abruptly to a reactive pattern at the onset of the second stage; B, the second stage continued with a normal CTG.

is relatively constant. Episodes of quiescence (inactivity) are associated with low FHR variability and non-reactivity (Figs 5.20 and 5.21). The fetus spends between 20 and 30 per cent of its time in periods of low activity during the day. Maternal activity does not seem to affect the different states of fetal activity or the FHR variability. It is, therefore, important to understand that low

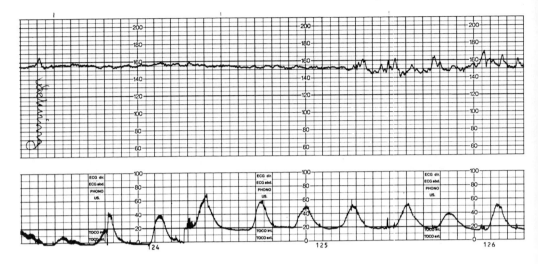

Fig. 5.24 External manipulation at the beginning of this antenatal record provoked only a small acceleration; 20 min later a spontaneous transition to the active behavioural state resulted in a normal, reactive, CTG.

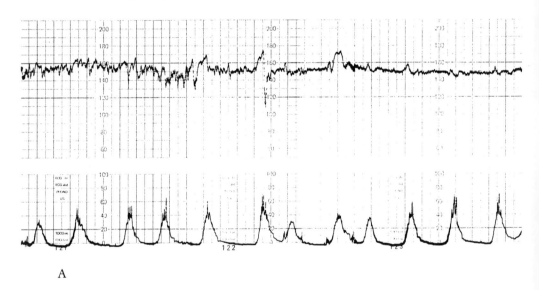

A

Fig. 5.25 (caption opposite).

variability and absence of accelerations may be related to the quiet episode of a fetal behavioural cycle when interpreting FHR traces (Figs 5.22 and 5.23). Such episodes seldom last longer than 40 min during late pregnancy or during labour.

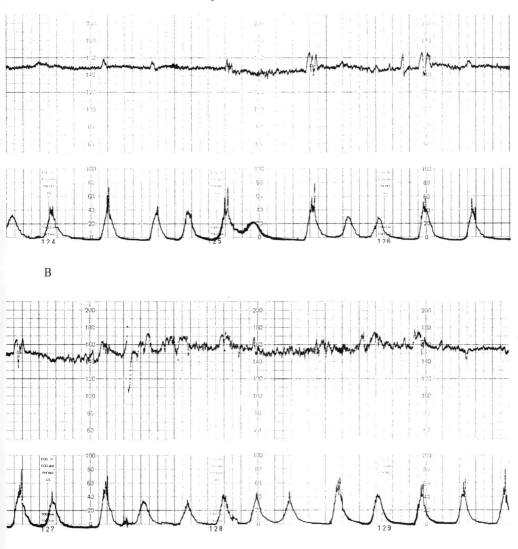

Fig. 5.25 An intrauterine pressure record and direct FHR during spontaneous labour. A, Low variability after mark 122 indicates quiet fetal behaviour; B, fetal quiescence continued for 90 min; C, normal variability returned after mark 127.

It sometimes happens that the behavioural state of the fetus changes from quiet to active state with external manipulation (Fig. 5.24). Vibroacoustic stimulation is an effective method of 'waking' the fetus and produces a transient tachycardia (state 4F) followed by persistence of the state 2F with its reassuring

FHR variability (see Fig. 7.9). This has been described as a means of reducing the number of false positive interpretations, which result from the low variability seen during short CTG records made during fetal quiescence. However, if a CTG recording made during fetal quiescence is continued, most healthy fetuses will change spontaneously to an active state within 40 min and low FHR variability will change to normal variability with accelerations. This pattern is characteristic of the normal fetus at term.

FHR variability may be modified by fetal breathing movements, although the relationship between behavioural state changes and episodes of fetal breathing movements is not clear. Fetal breathing movements occur less frequently during quiet episodes than during episodes of fetal activity.

Episodic changes of high and low FHR variability with cycles of active and quiet fetal behaviour also occur during labour. Episodes of low FHR variability last about 25 min (comparable to those in the antenatal period) although such episodes can last up to 90 min (Fig. 5.25). In many cases with such behavioural cycles during labour the band-width (amplitude) of the variability may be less than 5 bpm.

Interpretation of FHR variability

Normal changes in FHR variability, related to the cyclic alternation between quiet and active fetal states, should not be interpreted as 'reduction' or 'improvement' of variability. From a physiological point of view, phrases like

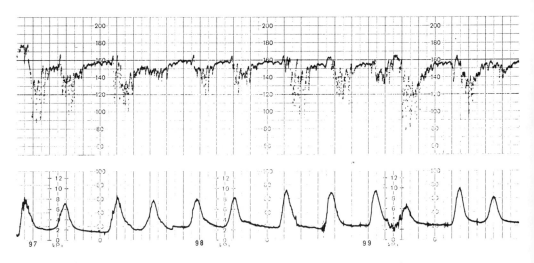

Fig. 5.26 An intrauterine pressure record and direct FHR during labour. The baseline rate is between 155 and 160 bpm. Baseline variability can only be assessed for periods of between 1 and 2 min between contractions.

Interpretation of FHR variability

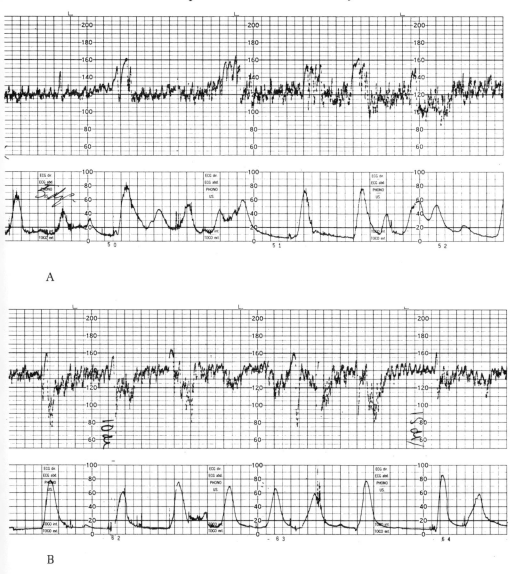

Fig. 5.27 (caption overleaf).

'the FHR variability was reduced for a 10-min period' is not significant. The significance of low FHR variability can only be assessed after an observation period of at least 40–50 min. In unclear cases, it is always judicious to prolong the recording and await the possibility of spontaneous improvement in variability related to a change in fetal behavioural state. If this does not occur soon after

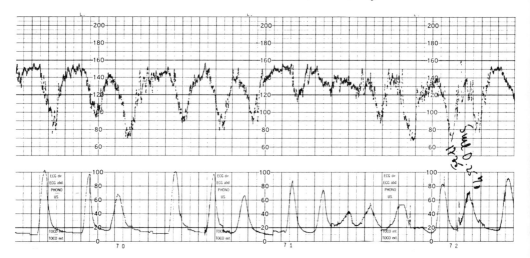

C

Fig. 5.27 An intrauterine pressure record and direct FHR during labour. A, The baseline rate is 120 bpm. FHR variability became high during some contractions whilst with others typical acceleration are seen; B, a slight rise in FHR occurred and fewer accelerations are seen with contractions. The FHR variability remained high and variable decelerations developed. C, 1 h later, just before onset of second stage, the FHR was between 140 and 150 bpm with typical variable decelerations. The baby was born in good clinical condition with the umbilical cord entangled three times around the neck. This pattern is described as a mixed cord compression pattern in the American literature.

40 min then an alternative explanation needs to be sought. Chronic hypoxia, associated with intrauterine growth retardation, is the most common pathological cause of low FHR variability with a 'non-reactive' FHR pattern.

During labour, the baseline variability can be assessed during the interval between contractions, which will be short when contractions are frequent (Fig. 5.26). This is especially important with traces showing pronounced variable decelerations (Chapter 9).

High FHR variability

High FHR variability, where the band width (amplitude) exceeds 25 bpm, can often be seen for short periods during labour and is especially obvious if the fetus has been stimulated externally. It is important to remember that all types of FHR variability may occur in a normal FHR trace for a short period of time. High FHR variability, especially when seen in combination with accelerations associated with uterine contractions, has been reported in relation to cord

Interpretation of FHR variability

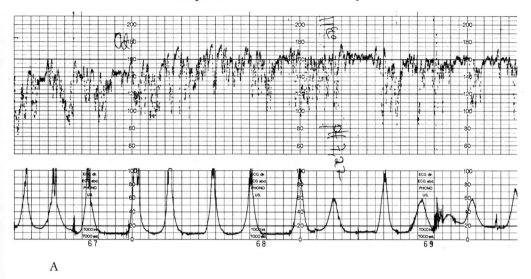

A

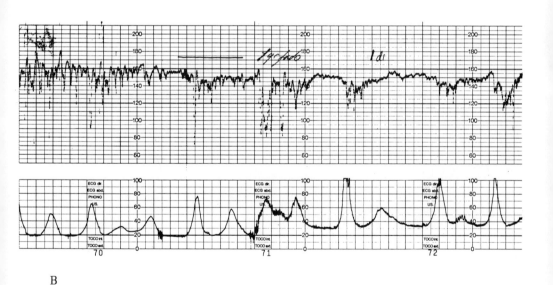

B

Fig. 5.28 Term labour in a primipara monitored by intrauterine pressure and FHR. A, The baseline rate was between 145 and 160 bpm with high FHR variability and variable (combined) decelerations. The fetal scalp pH was 7.27; B, variability returned to normal after the mother was put in the knee–chest position (at mark 71). Note that combined decelerations only occurred during contractions that generated high intrauterine pressure.

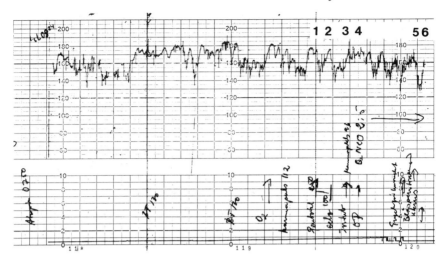

Fig. 5.29 A CTG made during anaesthesia for Caesarean section. The events were as follows: 1, thiopentone 250 mg; 2, suxamethonium 100 mg; 3, intubation; 4, skin incision; 5, uterine incision; 6, delivery of the fetus.

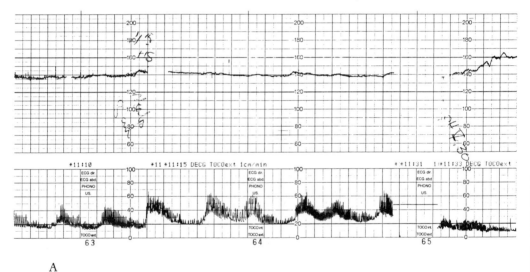

A

Fig. 5.30 (caption opposite).

compression and may represent acute hypoxia (Fig. 5.27). This accords with experimental animal data, which has shown an increase in FHR variability in response to acute hypoxia. Change of maternal position should be tried if cord compression is suspected. Figure 5.28 shows an example where the high variability disappeared when the mother was put in the knee–chest position.

Interpretation of FHR variability

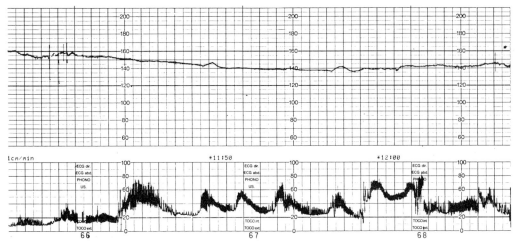

B

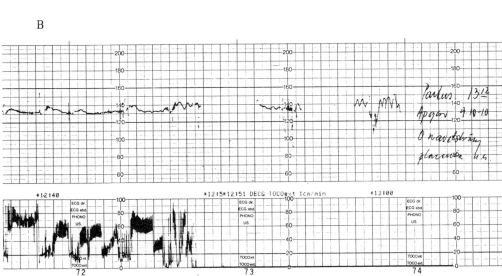

C

Fig. 5.30 A multipara at 40 weeks in spontaneous labour. The record shows the last 90 min of a 10-h labour. No medication was given. A, A short section of external record of FHR prior to amniotomy at 8 cm and application of a scalp electrode; the baseline FHR is 140 bpm. Note the absence of variability after commencement of the direct FHR (after mark 63) record indicating the degree of artefact on the external record. Fetal scalp blood pH was normal (7.3); B, the low variability was the indication for a second scalp pH (again normal); C, the CTG just prior to spontaneous delivery. Apgar scores were 9 and 10. This absence of variability was not associated with fetal hypoxia and acidosis.

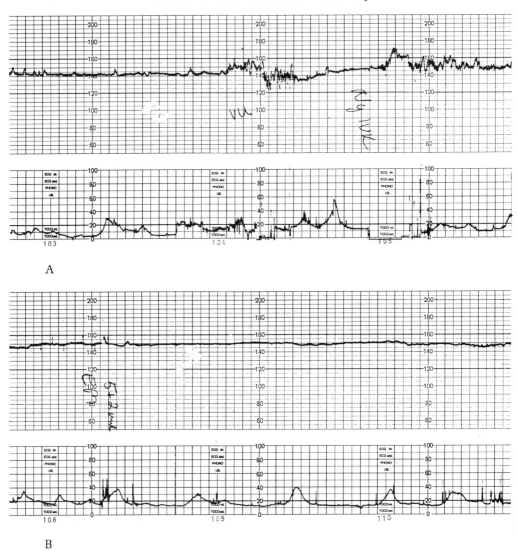

Fig. 5.31 (caption opposite).

Low FHR variability

A number of factors, apart from physiological ones, need to be considered if the FHR variability is less than 5 bpm:

1. Prematurity (see Chapter 13).
2. Tachycardia. A fast FHR (especially above 200 bpm) will cause low variability

Interpretation of FHR variability

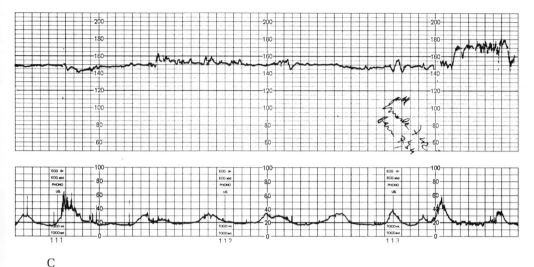

C

Fig. 5.31 A primipara at 40 weeks in spontaneous labour with a CTG showing low baseline variability but reactivity. An epidural had been given for analgesia prior to the first panel. A, A baseline FHR of 140 bpm with very low FHR. Reactivity occurred at mark 104 during insertion of an intrauterine catheter; B, the same case continued with virtual absence of variability; C, the FHR showed variability at the time of a fetal blood sample (mark 112) and the pH was normal (7.34).

because reduction in pulse interval duration produces less variation in interval differences between beats.

3. Drugs (see Chapter 16).
4. General anaesthesia. Many drugs cross the placenta and can influence the fetus (e.g. anaesthesia for Caesarean section (Fig. 5.29).
5. Fetal malformation. Heart malformations, in particular, are associated with low FHR variability but malformations of the central nervous system have also been implicated (see Chapter 6).
6. Hypoxia. This is an important pathological cause of low FHR variability. The combination of low variability and tachycardia is discussed in Chapter 6. Studies during labour have shown that fetal acidosis appears in about half of all cases with low variability lasting longer than 180 min. Low variability is a feature of traces with a terminal pattern (see Chapter 15). Cerebral damage during the antepartum period may also result in very low or absent variability. For example, multiple placental infarcts may result in chronic injury (see Fig. 15.4) and cases have been reported where a sudden loss of fetal movements, associated with very low FHR variability, prior to labour have indicated brain death, as evidenced by neonatal condition after delivery (usually by Caesarean section).

However, low FHR variability can sometimes be seen for several hours without signs of fetal distress at birth (Figs 5.30 and 5.31). Figure 5.30 illustrates such a case where very low FHR variability was present for 10 h during labour. However, normal FHR variability was seen during vaginal examinations, insertion of an intrauterine catheter, and with fetal scalp blood sampling. The FHR responses to these manoeuvres indicated that fetal hypoxia was not present.

Summary

A full understanding of the control of FHR and influences on FHR variability is necessary for appropriate interpretation of the CTG. It is seldom possible to make correct deductions about fetal well-being from the FHR record without full knowledge of the clinical circumstances at the time of monitoring. Retrospective interpretation is especially difficult if an adequate record of the clinical situation has not been made. With increasing use of CTG records to substantiate claims of malpractice it is imperative that a complete clinical record accompany any note regarding interpretation of the FHR, including an opinion and a plan for subsequent management.

6
Baseline fetal heart rate

Introduction

The average rate around which the FHR continuously varies is considered to be the baseline FHR. It is the rate obtained by auscultation at any given point in time and it can be estimated from the continuous FHR record between accelerations and decelerations (Fig. 6.1). The physiological state of the fetus (active or quiet) should be known for optimal interpretation, as the rate may be higher during fetal activity. Similarly, any transient influence of uterine activity on the FHR during labour must be assessed before the baseline rate can be accurately determined (Figs 6.2 and 6.3). It may not be possible to record the FHR in cases of arrhythmia (see Fig. 2.3) and it is difficult to know what the 'normal' rate is in such circumstances. It is important to recognize the possibility that the external (ultrasound) transducer may create artefacts, particularly when the FHR is high (above 180 bpm) or low (less than 90 bpm), and it is sensible to check the real FHR with a stethoscope or real-time ultrasound imaging in such cases (see Figs 2.7 and 2.10).

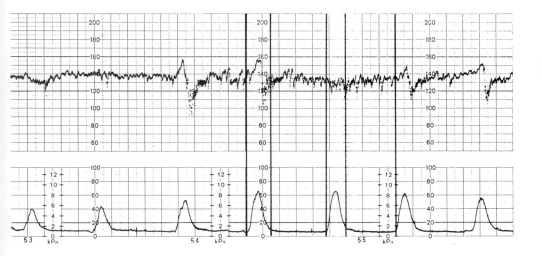

Fig. 6.1. Baseline FHR should be evaluated between uterine contractions.

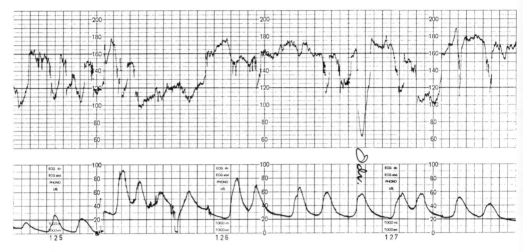

Fig. 6.2 An external record of uterine activity showing overstimulation with oxytocin. Baseline FHR (between 160 and 170 bpm) is difficult to identify due to the frequent variable decelerations.

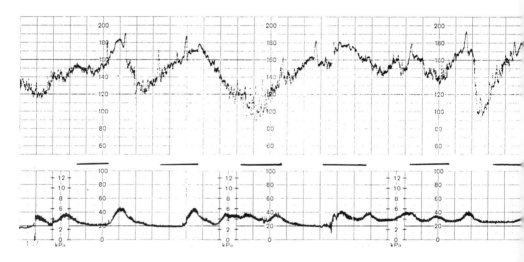

Fig. 6.3. Baseline FHR 'wandering' slowly between contractions and difficult to classify — probably between 170 and 180 bpm.

Normal range

Traditionally, the normal FHR has been considered to be within the range of 120–160 bpm. Tachycardia is mild between 160 and 180 bpm and severe above

180 bpm. Similarly, bradycardia is mild between 120 and 100 bpm and severe below 100 bpm. Recently, the International Federation of Gynaecology and Obstetrics (FIGO) has recommended that the normal limits to baseline FHR be altered to between 110 and 150 bpm.

Changes in FHR may occur slowly or rapidly. A tachycardia in labour usually develops slowly in contrast to most bradycardias below 100 bpm, which occur rapidly. However, baseline bradycardia (between 100 and 110 bpm) may also develop slowly. In an unselected population of 2566 women in labour, continuously monitored during the first stage of labour, tachycardia was seen in 6.2 per cent of labours and bradycardia was seen in 8.5 per cent during an observation period of at least 30 min. In a low-risk population of 4278 women, tachycardia was less frequent (4.8 per cent) but the frequency of bradycardia (7.7 per cent) was similar to that in an unselected population.

The risk of fetal heart decompensation increases if the rate exceeds a rate of 200–220 bpm. A mild tachycardia in the range of 150–170 bpm, developing during labour, may be a sign of early fetal hypoxia and catecholamine release. Consequently, late decelerations occur more commonly with a fetal tachycardia. Reduced variability and late, combined, or variable decelerations are more likely to indicate fetal hypoxaemia and acidaemia if they occur together with a tachycardia (Fig. 6.4).

On the other hand, an FHR between 100 and 110 bpm, if reactive with normal FHR variability, is seldom associated with fetal distress (Figs 6.5 and 6.6).

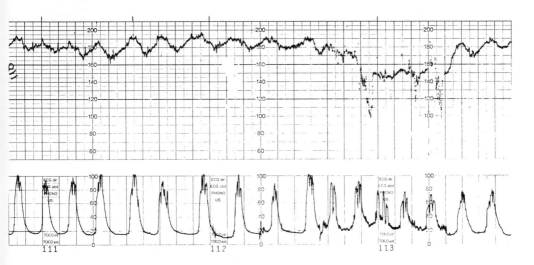

Fig. 6.4. An FHR tachycardia (190 bpm) with recurrent, shallow, late decelerations and low variability. Intrauterine pressure record of uterine activity. Scalp blood pH = 7.15 (at the beginning of the record). The women was delivered by Caesarean section, Apgar scores were 3, 8, and 10; the cord artery pH was 7.04.

86   Baseline fetal heart rate

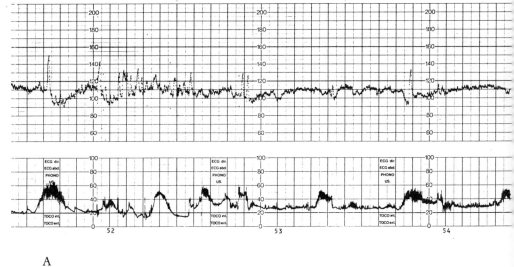

A

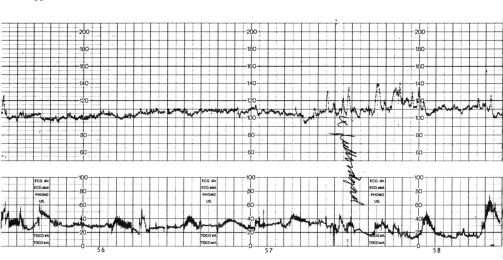

B

Fig. 6.5 (caption opposite).

A terminal FHR pattern with this rate is usually preceded by abnormal FHR changes, such as severe FHR decelerations and absence of FHR variability (see Chapter 15). In a study involving 218 women with an FHR between 100 and 120 bpm, in the first stage of labour only two newborns had an Apgar score below 7 at 5 min.

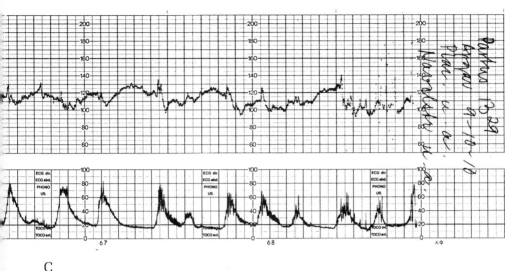

C

Fig. 6.5 Baseline bradycardia in a primipara at 39 weeks in spontaneous labour. External tocography. A, the baseline rate is 110 bpm with normal FHR variability and reactivity and no decelerations; B, 30 min later, a quiet fetal behavioural state is seen in the first part of this record. The fetus responded with accelerations during the vaginal examination (between marks 57 and 58); C, record during the second stage with a mild bradycardia as before. Apgar scores were 9 and 10.

Tachycardia

Fetal origin

Hypoxia

Although tachycardia may be an early sign of fetal hypoxia, features such as normal FHR variability, reactivity, and absence of decelerations, are reassuring (Fig. 6.7). The situation is different when the tachycardia occurs with other FHR abnormalities. A combination of tachycardia and reduced FHR variability, without accelerations, in the first stage of labour merits further investigation (Fig. 6.8). Tachycardia and recurrent late decelerations are related to fetal hypoxia, particularly if the variability is very low (Figs 6.9 and 6.10).

Compensatory tachycardia

This is often seen with intense fetal activity (Fig. 6.11) and is normal. Other causes include anaemia from rhesus isoimmunization or, rarely, fetal bleeding from accidental puncture of placental vessels at amniocentesis (Fig. 6.12) or trauma to vasa praevia at membrane rupture during labour (see Fig. 11.6). After

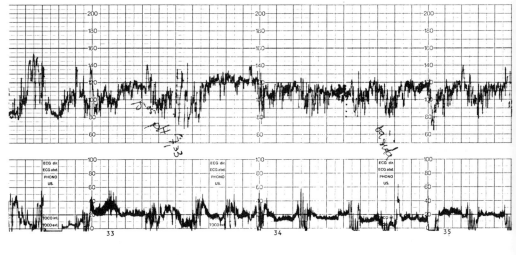

A

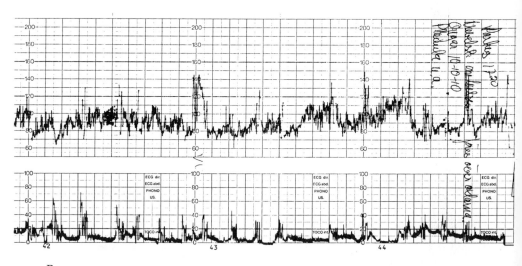

B

Fig. 6.6 Spontaneous labour in a primipara at 40 weeks. External tocography. A, The FHR 135 min prior to delivery shows a baseline bradycardia of 115 bpm and high FHR variability. Variability increased during contractions, which are not well recorded. Fetal scalp blood pH was 7.33; B, the same FHR pattern continued during the second stage. Apgar scores were 10 and 10. The umbilical cord was around the fetal neck at delivery.

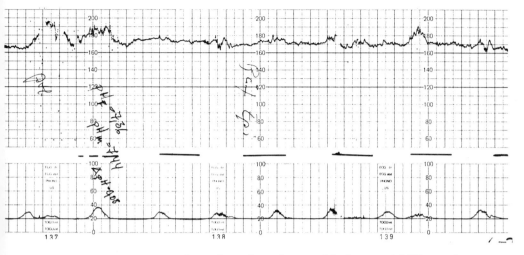

Fig. 6.7 A reactive tachycardia of 170 bpm with low variability and no decelerations. Normal scalp blood pH = 7.36.

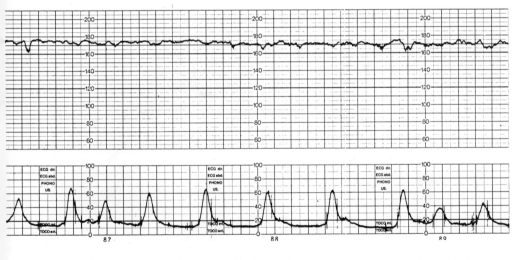

Fig. 6.8 A primipara at 39 weeks in prolonged labour after rupture of the membranes for 24 h. The maternal temperature was normal. The CTG shows a fetal tachycardia of 170 bpm with low/absent variability and no reactivity or decelerations. Scalp blood pH was 7.22. A Caesarean section was performed; the Apgar scores were 3 and 9.

a prolonged deceleration a transient period of tachycardia is often seen, possibly indicating a response to further catecholamine production secondary to the hypoxia which produced the initial deceleration (see Fig. 11.9).

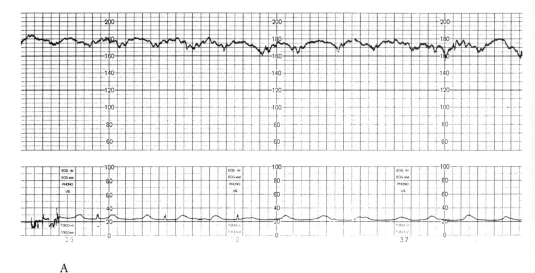

A

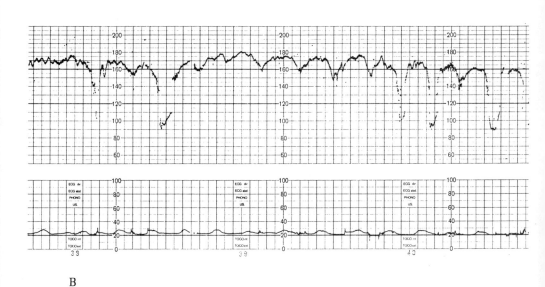

B

Fig. 6.9 A multipara at 37 weeks with pre-eclampsia and vaginal bleeding in spontaneous labour. A, The baseline FHR is between 175 and 180 bpm with low FHR variability and no reactivity; there are shallow, late, repetitive decelerations; B, the late decelerations have become more pronounced. Delivery was by Caesarean section; Apgar scores were 5 and 9. The cord artery pH was 7.26 and the cord venous pH was 7.33. A retroplacental blood clot confirmed an abruption of the placenta.

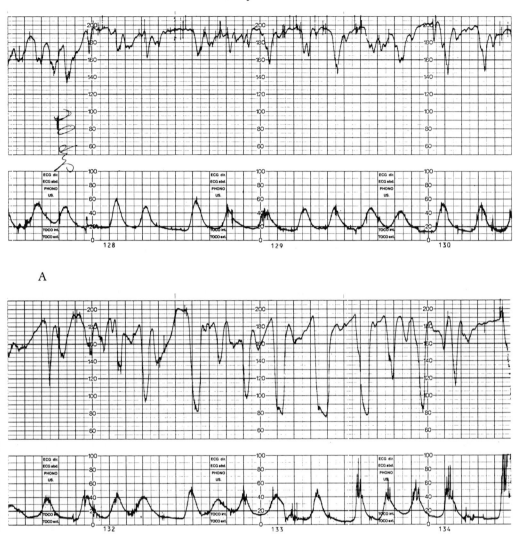

Fig. 6.10 (caption overleaf).

Fetal behavioural state and tachycardia

A sudden change of FHR occurs spontaneously after a period of quiet sleep or may be induced by pain or stimulation tests, such as vibroacoustic stimulation (Fig. 6.13). In general, the FHR returns to normal within 20–30 min. Normal variability and accelerations are usually seen.

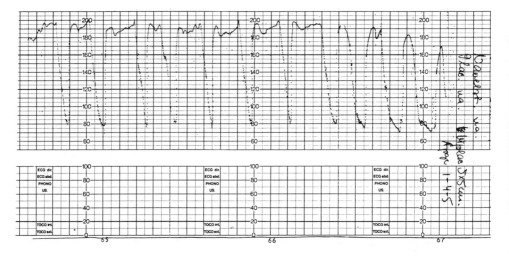

C

Fig. 6.10 A primipara at 38 weeks with a breech presentation in spontaneous labour augmented by oxytocin infusion and monitored by intrauterine pressure. Abnormal features appeared after several hours of labour with a normal FHR. A, 80 min prior to delivery the baseline rose to 190 bpm with absent short-term variability. Variable and combined decelerations can be seen. The cervix was fully dilated with the breech just below the ischial spines; B, 10 min later the tachycardia and absent short-term variability were still present. Pronounced combined (variable and late) decelerations had developed; C, the last 30 min of labour prior to assisted breech delivery showing continuation of a worrying FHR. Apgar scores were 1, 4, and 5, indicating severe clinical depression of the newborn. Intervention was appropriate at panel A.

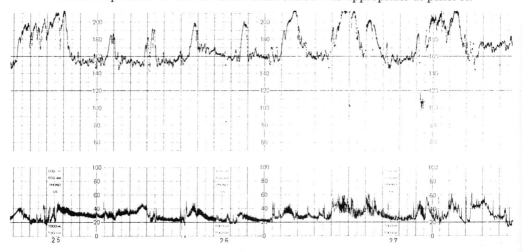

Fig. 6.11 Frequent fetal movements with large sporadic and periodic FHR accelerations.

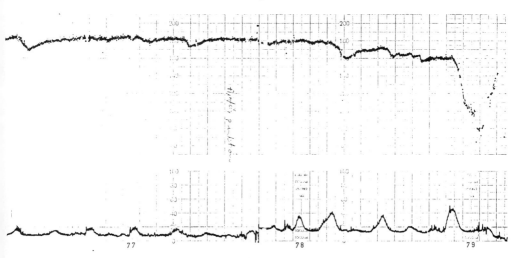

Fig. 6.12 Fetal haemorrhage (following puncture of a placental vessel during amniocentesis at 36 weeks) resulting in a tachycardia of 180 bpm with absent short-term and long-term variability (the external recording gives an erroneous impression of variability). Occasional late decelerations can be seen. At Caesarean section the bleeding was verified. The fetus was in mild hypovolaemic shock with Apgar scores of 4, 6, and 9.

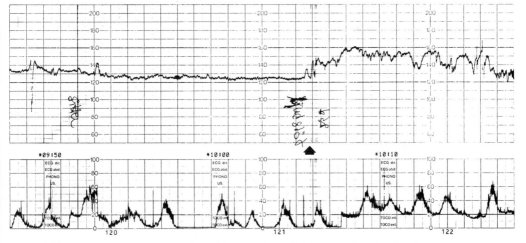

Fig. 6.13 A transient tachycardia and accelerations after fetal vibroacoustic stimulation (at the arrow).

Prematurity

From a baseline FHR of 175 bpm in early pregnancy the rate gradually falls until 30 to 32 weeks gestation, after which it remains reasonably stable. It may be difficult to differentiate between a benign mild tachycardia (around 150 bpm)

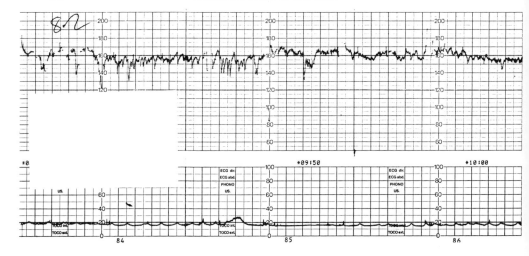

Fig. 6.14 An antepartum CTG at 29 weeks showing a slight tachycardia of 165 bpm with normal variability. The small decelerations, particularly in the first part of the tracing, are also a normal feature of early third trimester FHR records.

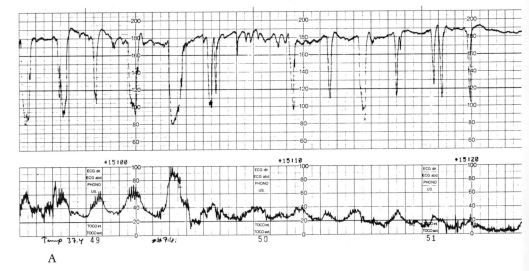

A

Fig. 6.15 (caption opposite).

in the preterm period and an increased FHR due to other causes, such as fetal infection (Figs 6.14 and 6.15).

Fetal heart arrhythmias

Tachyarrhythmia and atrial flutter may be associated with severe fetal tachycardia.

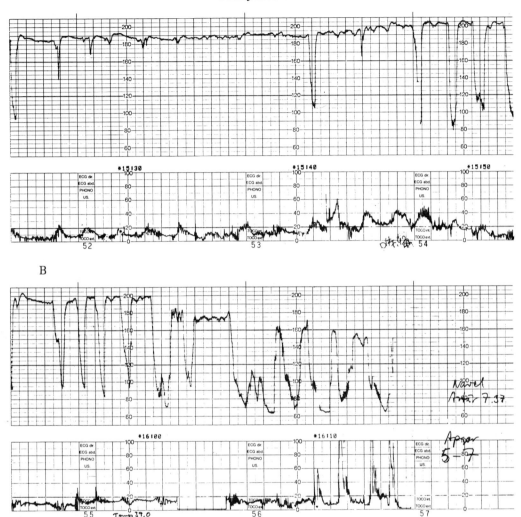

Fig. 6.15 A multipara at 29 weeks in spontaneous labour after ruptured membranes for 12 weeks. External tocography. The records show the last 85 min of the labour. A, A tachycardia of between 180 and 185 bpm with absent short-term variability and low long-term variability. Variable decelerations of relatively short duration are present and some are followed by 'overshoot' of the FHR. The maternal temperature was 37.4 °C; scalp blood pH was 7.41; B, the tachycardia increased to between 190 and 200 bpm with no change in variability. Occasional variable decelerations persisted. Scalp blood pH was 7.42; C, a maternal fever (39 °C) developed and delivery occurred 15 min later. Birthweight was 1250 g; Apgar scores were 5 and 7; the cord artery pH was 7.37. Although the fetus was in good condition at birth in this case there was a high risk of fetal acidosis associated with such a FHR record.

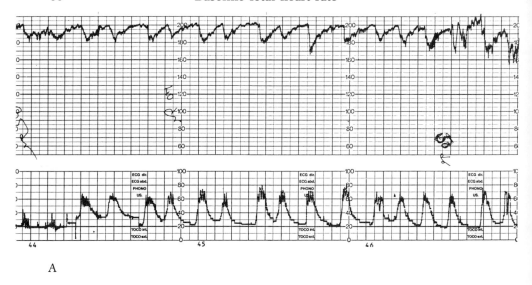

A

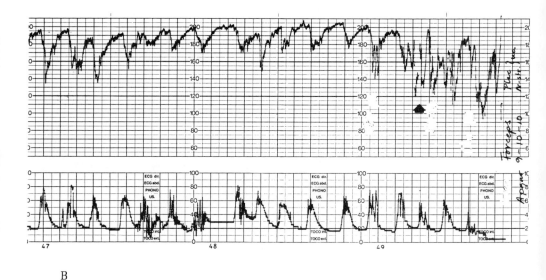

B

Fig. 6.16 The last 60 min of a labour lasting 15 h in a primipara. External tocography. A, The baseline FHR was 200 bpm with low variability and early decelerations; B, the pattern continued until application of forceps (at the arrow) for delivery. Apgar scores were 9 and 10.

Tachycardia

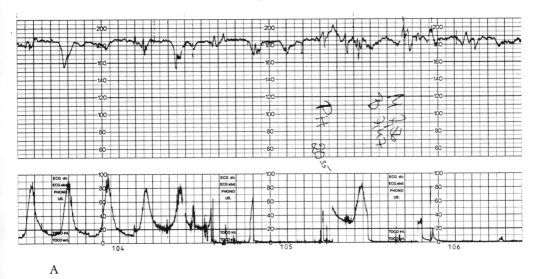

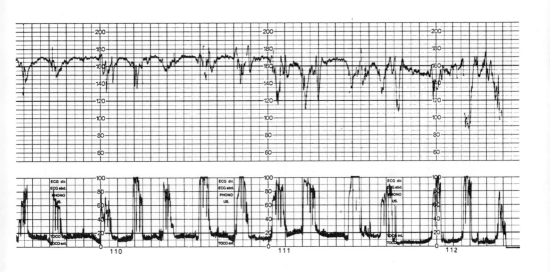

Fig. 6.17 A primipara in spontaneous labour at term with intrauterine pressure monitoring. A, The baseline FHR was 185 bpm with low FHR variability and reactivity. Fetal scalp blood pH was 7.47 and maternal venous blood pH was 7.56; B, the second stage of labour 40 min later showed mild, variable, decelerations. The Apgar scores were 9 and 10. Maternal hyperventilation had produced a respiratory alkalosis (high pH value) in maternal blood.

Maternal origin

Fever

The maternal temperature should be monitored in cases with tachycardia. We found maternal pyrexia (above 38 °C) in 20 per cent of all cases with a fetal tachycardia (above 160 bpm) in labour. It has been stated that baseline tachycardia may occur before a rise in maternal temperature but our findings do not concur with this. Sixty-nine (2.3 per cent) patients developed a fever (equal to or greater than 38 °C) in the first stage of labour during 1989 in Lund. The maternal temperature was measured at regular intervals (every 2–4 h). Fetal tachycardia was recorded in 57 of these cases and only 13 had a tachycardia before a rise in the temperature was recorded. The FHR may return to normal if the maternal fever is reduced. Maternal fever is not uncommon in cases with prolonged labour (Fig. 6.16).

Epidural analgesia in labour has been associated with maternal pyrexia, probably as a consequence of a disturbance to maternal thermoregulatory mechanisms. As a result, fetal baseline tachycardia may develop.

Increased sympathetic tone

Maternal distress in women with a lot of pain often results in fetal tachycardia (Fig. 6.17). Adequate analgesia will usually produce a reduction in FHR.

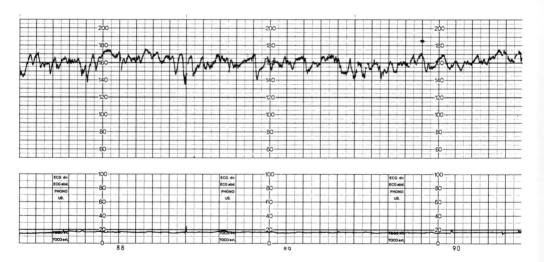

Fig. 6.18 A primipara at 30 weeks in preterm labour. An infusion of terbutaline (15 μg/min) inhibited uterine activity. The baseline FHR was between 160 and 165 bpm with normal FHR variability.

Pharmacological

Treatment of preterm labour with a β-receptor agonist will produce a maternal tachycardia of up to 120 bpm. A concomitant fetal tachycardia above 150 bpm is seen in about 80 per cent of cases but is seldom severe (Fig. 6.18). Such fetuses usually have a higher FHR anyway because of the earlier gestation.

Administration of atropine to the mother has been shown to induce fetal tachycardia.

Other maternal causes

Fetal tachycardia associated with maternal hyperthyroidism would suggest an effect of the disease on fetal thyroid function.

Bradycardia

Fetal origin

Head compression

Sustained vagal stimulation may result from raised intracranial pressure secondary to pressure on the fetal head during labour. It may occur with rapid progression of labour with rapid descent of the fetal head through the birth canal (Fig. 6.19).

Regional analgesia

Bradycardia up to 20 min in duration occasionally follows administration of a paracervical or epidural block, especially if there is maternal hypotension (Fig. 6.20).

Analgesics

Narcotic drugs, such as pethidine, may cause a slight reduction in baseline rate (of between 10 and 20 bpm), which may be associated with a reduction in fetal activity (Fig. 6.21).

Atrioventricular block

See Fig. 6.29.

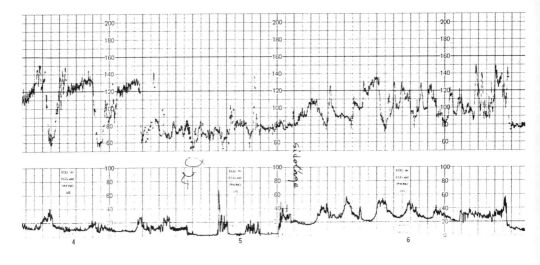

A

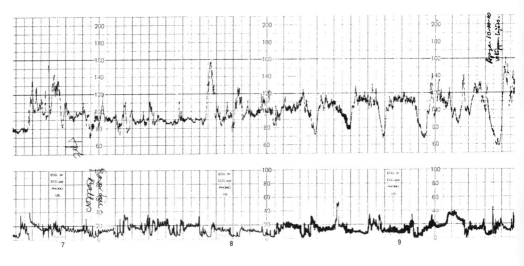

B

Fig. 6.19 A primipara at term in spontaneous labour with an occiput–posterior fetal position. The last hour of labour is shown. A, Variable decelerations (at mark 4) were followed by a bradycardia lasting 8 min. The baseline FHR thereafter was between 90 and 110 bpm; B, continuation of the FHR until just before delivery. Apgar scores were 10 and 10. The FHR had variability and reactivity, and was therefore reassuring.

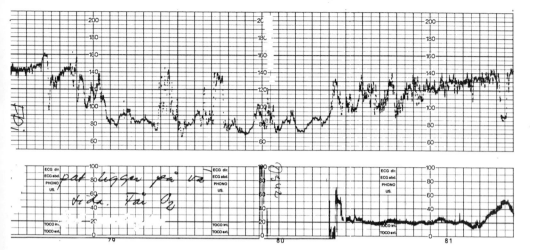

Fig. 6.20 An FHR bradycardia lasting 15 min following an epidural for analgesia.

Hypoxia

Acute hypoxia during contractions may invoke a transient, chemoreceptor-mediated, bradycardia (deceleration), but these may be tolerated for some time before fetal reserve and adaptation diminish. The bradycardia associated with significant hypoxia and acidaemia is usually pronounced and accompanied by other abnormal changes, such as reduced baseline variability and/or variable or late decelerations (Figs 6.22 and 6.23).

Idiopathic

A baseline rate between 100 and 110 bpm can occasionally be found during the whole of the first stage of labour without any adverse reason. When reactive, with accelerations and normal variability, this FHR pattern is not associated with hypoxia (Fig. 6.24).

Maternal factors

Excessive uterine activity

This is one of the most common causes of acute fetal bradycardia and is usually iatrogenic. Uterine sensitivity to oxytocin varies from individual to individual, and even a slow infusion may provoke hypertonic uterine activity (Fig. 6.25). If possible, oxytocin should be given by an infusion pump to avoid overdose, and the dose should be reduced once labour progress has been restored.

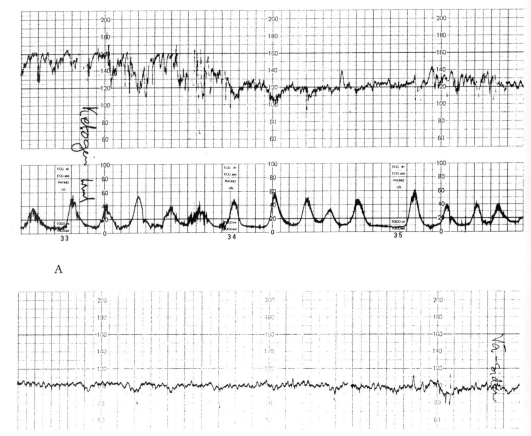

Fig. 6.21 A narcotic analgesic was given at mark 33. The FHR baseline stabilized at 120 bpm after 10 min; B, the FHR remained unreactive 1 h later.

Hypovolaemia

This may occur after a prolonged labour of several hours. The bradycardia is often mild.

Bradycardia

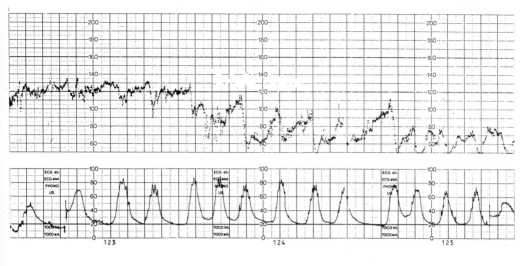

A

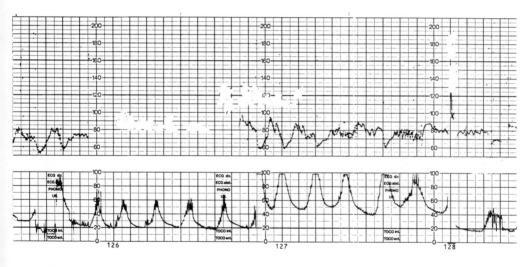

B

Fig. 6.22 A primipara at 39 weeks with a breech presentation in spontaneous labour. The last 60 min of labour are shown. A, During the first 10 min the FHR pattern was normal. The rate then fell as the second stage of labour was reached. Initially, the FHR variability remained normal with variable and combined decelerations; B, the scalp electrode fell off for 10 min after which the baseline FHR can be seen as 80 bpm with decreasing variability. The infant was depressed, with Apgar scores of 2, 6, and 7. This bradycardia was allowed to persist for too long.

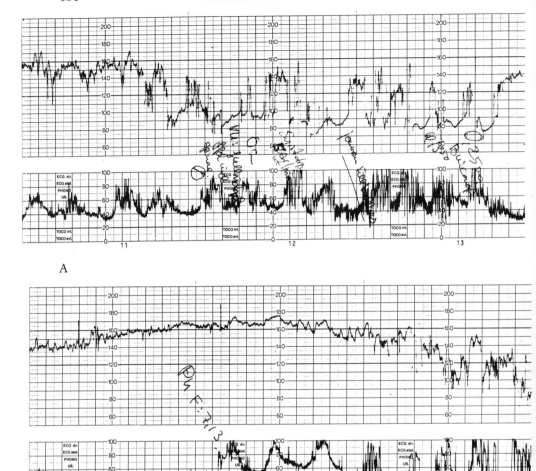

Fig. 6.23 A primipara at 39 weeks in spontaneous labour with an oxytocin infusion for augmentation of labour. A, A normal, reactive, FHR pattern for 10 min, after which there is a sudden onset bradycardia with FHR below 100 bpm. Variability is absent, particularly between the rises in FHR, which nearly reach the previous baseline. The oxytocin infusion was stopped at mark 12 but the bradycardia continued, interrupted by short attempts to return to the normal baseline. Terbutaline (0.25 mg i.v.) was given at mark 13; B, the FHR responded to the uterine relaxation by developing a mild compensatory tachycardia after the injection. Scalp blood pH was 7.13. A mid-forceps delivery was performed at the end of panel B. Apgar scores were 7 and 10 and the cord artery pH was 7.15. Fetal acidosis will usually develop after bradycardia lasting for 20 min or more.

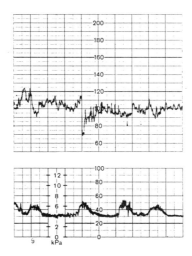

Fig. 6.24 An uncomplicated bradycardia with a baseline rate of 100 bpm.

Hypotension

The most common cause of maternal hypotension is aortocaval compression (the vena cava syndrome) (Fig. 6.26). This occurs when the mother lies supine and is often seen at the time of a vaginal examination.

Fetal heart arrhythmia

There are special problems with the interpretation of the FHR in cases of fetal arrhythmia because the monitors are not always capable of recording irregular signals. Four main types of arrhythmia occur: (i) supraventricular tachycardia; (ii) atrial flutter; (iii) atrioventricular block; and (iv) extrasystoles.

With supraventricular tachycardia, the fetal heart rate can reach a rate of 200 bpm or more. The tachycardia is often paroxysmal, but may be continuous. Reduced baseline variability is common. The risk of fetal heart failure is increased and frequent ultrasound examinations are indicated to rule out pericardial effusion, heart dilatation, liver enlargement, or ascites. If heart failure is suspected, treatment with digitalis *in utero* may be considered if delivery is not considered appropriate. Fetal heart malformations are found in 5 to 10 per cent of cases. Benign types of tachycardia are those that change into normal rhythm during pregnancy or in labour (Fig. 6.27).

In atrial flutter, the auricles of the fetal heart have a frequency of 200–300 bpm, which cannot be recorded with a monitor. The ventricular response is either arrhythmia or some kind of block (often 2 to 1). Figure 6.28 shows such a case with an auricular frequency of 372 bpm and a 2 to 1 block

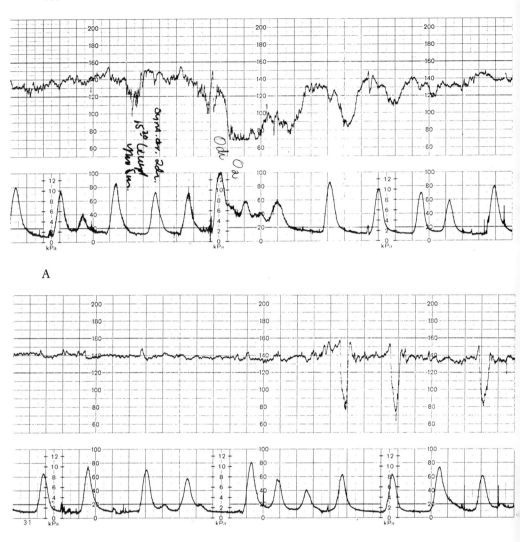

Fig. 6.25 A primipara at term in spontaneous labour augmented with a low dose of oxytocin (2 mU/min). Internal tocography. A, Abnormal uterine activity lasting at least 5 min was associated with a bradycardia below 100 bpm. The oxytocin infusion was stopped and the FHR pattern became normal after three combined decelerations at the end of the panel; B, a normal FHR pattern except for three small variable decelerations. A compensatory tachycardia after the prolonged deceleration was seen not in this case. There was no indication for fetal blood sampling given the immediate recovery in FHR.

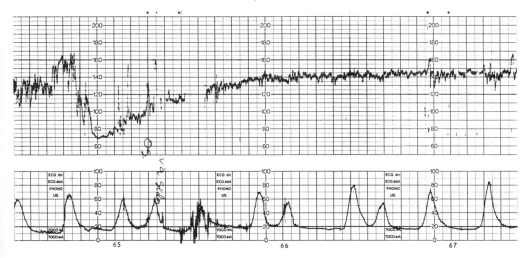

Fig. 6.26 Spontaneous labour with intrauterine pressure measurements. At mark 65 a bradycardia occurs in association with the supine position. Recovery occurred after correction of the maternal position.

with ventricular frequency of 186 bpm. This produces a 'fixed', straight-line FHR. As the CTG record gives no information about fetal condition, repetitive fetal blood samples to monitor fetal acid–base balance is necessary. In this specific case the pH values during labour were normal and the baby was born with Apgar scores of 8 at 1 min and 9 at 5 min. Umbilical cord artery pH at delivery was 7.28. Figure 6.28(c) shows the fetal ECG soon after delivery with doubled P waves indicative of atrial flutter. This type of arrhythmia may be treated with digitalis in utero, especially with high ventricular frequencies, which increase the risk of heart failure. Fetal heart malformations may also be present in these cases and should be ruled out in the antenatal period.

Fetal atrioventricular block (AV block) is uncommon. The FHR pattern shows a pronounced bradycardia of 60–80 bpm and a 'fixed' baseline with absent variability. Fetal heart malformations are common (30–50 per cent). The condition may be related to maternal autoimmune diseases. Figure 6.29 shows a case of total AV block. The fetal heart rate is fixed at around 70 bpm. Repeated fetal pH measurements during labour showed normal values (7.39, 7.33). Not even during the second stage did changes in the FHR become remarkable. The Admission Test (panel A) 30 min before panel B presents a reactive pattern. It is plausible that the record at panel A is of maternal origin. It is recommended that these patients should be referred to a specialist unit with facilities for neonatal intensive care and expertise in paediatric cardiology and thoracic surgery.

108 Baseline fetal heart rate

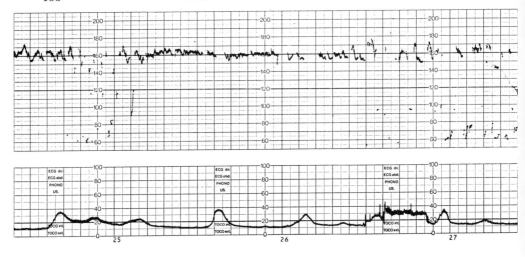

A

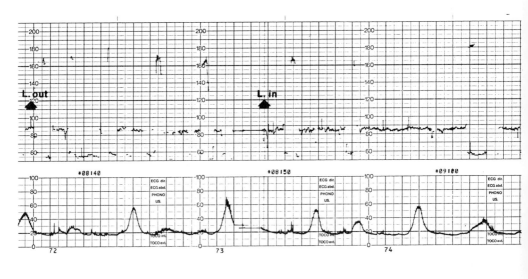

B

Fig. 6.27 A multipara at 32 weeks with an irregular FHR on auscultation. A, Periods of arrhythmia; B, the next day showed a more severe arrhythmia with supraventricular extrasystoles in bigeminal and trigeminal rhythm. The logic system of the monitor (HP8040 A) was switched off at L.out (mark 72) and switched on at L.in (mark 73). An ultrasonic scan showed a normally formed heart and no signs of decompensation. Doppler investigation of the aorta and umbilical artery blood flow showed normal values; 1 week later the fetal heart action was normal.

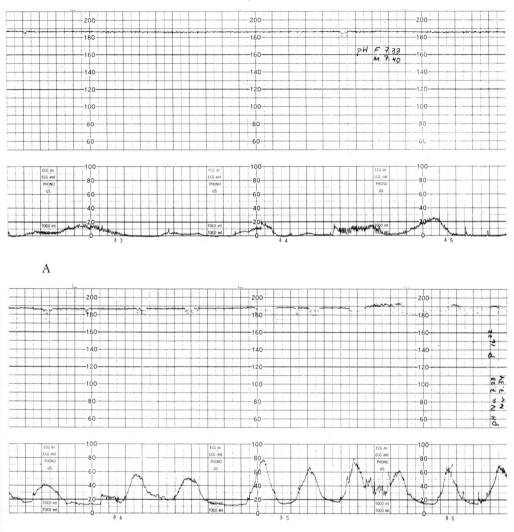

Fig. 6.28 (caption overleaf).

Supraventricular or ventricular extrasystoles are occasionally found in the general population and are not related to fetal heart malformations. They may sometimes be seen as momentary interruptions in the CTG record (see Fig. 2.2). If the extrasystoles are sparse, the FHR pattern may look normal. If they occur more frequently then the record may become difficult to interpret (see Fig. 2.3). Fetal extrasystoles often disappear spontaneously.

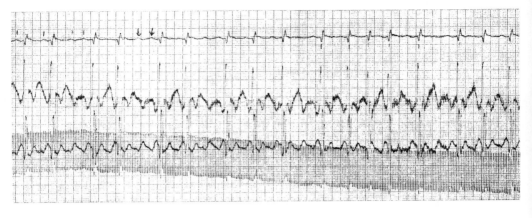

C

Fig. 6.28 A, Atrial flutter with a fixed FHR baseline. Fetal scalp blood pH was normal (7.33); B, the last 30 min before delivery. Cord artery pH was 7.28 and cord venous pH was 7.34; C, the ECG from the newborn soon after birth showed double P waves (arrows).

Summary

In the absence of behavioural state change or maternal pyrexia, fetal tachycardia should be regarded as an indication of catecholamine secretion in response to stress. Underlying maternal and fetal causes should be looked for and hypoxia should be excluded. Similarly, an unexpected fetal bradycardia of sudden onset is likely to represent acute fetal hypoxia.

The rate of fetal distress in labour may not be substantially increased in cases with an arrhythmia, although the FHR pattern may look dramatic and cause much anxiety. Facilities for fetal blood gas analysis are essential for fetal supervision in labour. Ultrasound examinations of the fetal heart and blood flow velocity examinations in fetal vessels are of great value for the obstetrician for giving balanced (and often reassuring) information to the parents.

Summary

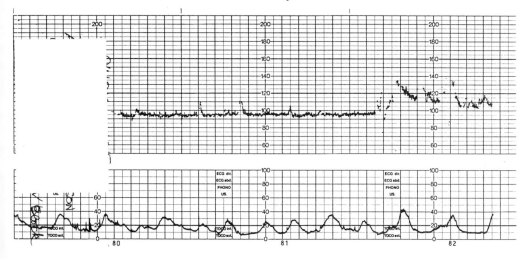

A

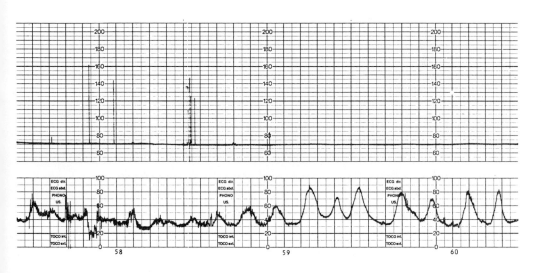

B

Fig. 6.29 (caption overleaf).

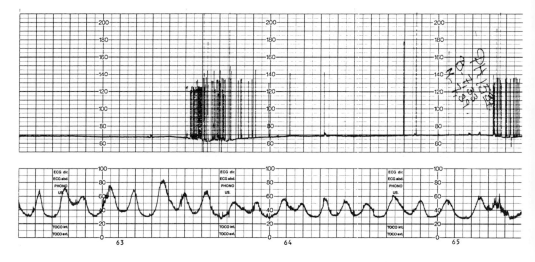

C

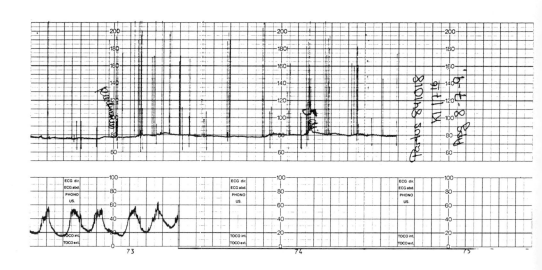

D

Fig. 6.29 An example of atrioventricular block during labour at term. A, The Admission Test (the maternal heart rhythm was probably recorded); B, a scalp electrode was applied and showed a baseline of 70 bpm during the first stage of labour; C, 30 min later, the scalp blood pH was 7.33. Another pH taken 40 min later was also normal; D, the second stage record. Apgar scores were 8, 7, and 9. The baby had a lethal congenital heart malformation.

7
Accelerations

Introduction

An acceleration is a transient increase in FHR that returns to the baseline rate. The baseline rate is usually taken as the mean value between upper and lower limits of variability. The most widely accepted definition is an abrupt rise (amplitude) of at least 15 bpm lasting at least 15 s (Fig. 7.1) but there are occasions, especially if baseline variability is low, when discrete and transient rises with an amplitude of 10 bpm may be considered to be accelerations (Fig. 7.2).

Mechanism of FHR accelerations

FHR accelerations usually accompany fetal body movements and are the most important parameter by which fetal well-being may be recognized from the CTG. They may occur with movements that are not perceived by the mother but some

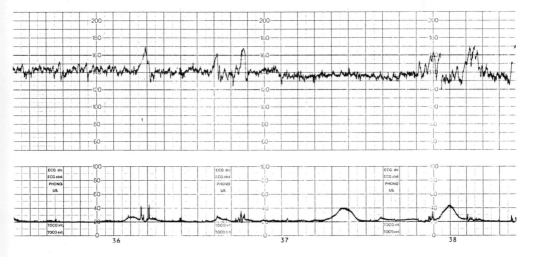

Fig. 7.1 An antepartum CTG with sporadic accelerations associated with fetal movements, indicated on the toco trace.

114   Accelerations

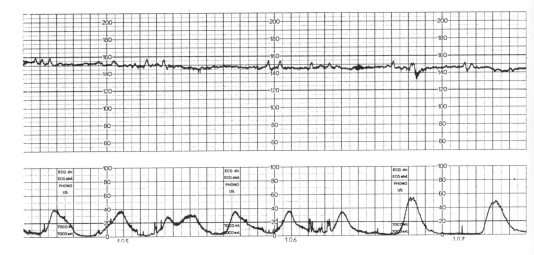

Fig. 7.2 Small (low amplitude) accelerations associated with fetal movements.

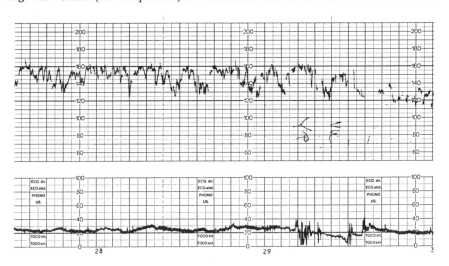

Fig. 7.3 Transient elevation of the baseline FHR during prolonged fetal activity. It may be difficult to be sure of the baseline FHR until the fetal activity ceases.

fetal movements, usually small ones such as limb movements seen by ultrasound, are not always associated with accelerations. Absence of FHR accelerations in the presence of perceived fetal movements should be regarded as a sign of early fetal hypoxaemia.

Sporadic accelerations are related to gross fetal body movements, which have a duration of more than 3 s. If the fetus remains very active for a period of time then accelerations of the FHR may be so frequent and repetitive that they merge to become a raised baseline (Fig. 7.3). If this new baseline is above 150 bpm (the upper limit of normal) during fetal activity it may be considered a transient tachycardia. The recording should be continued until the fetal activity settles so that the true baseline rate can be determined. When frequent accelerations resemble high FHR variability, differentiation of the baseline rate can be difficult (Fig. 7.4).

Pronounced accelerations, which merge during fetal activity in labour may also produce an apparent, transient elevation of the baseline. Occasionally, the FHR returns to the real baseline with contractions giving the appearance of late decelerations (Fig. 7.5). Such accelerations often have a 'step-ladder' or 'roofing tile' appearance, which differs from the more regular 'serrated' appearance of baseline variability. It is appropriate to adopt a conservative (wait-and-see) policy, especially during fetal hyperactivity states, to determine the normal baseline rate, which will usually reappear spontaneously (Figs 7.5 and 7.6).

Fetal activity during labour can often be seen on the toco trace of the CTG between uterine contractions (Figs 7.7 and 7.8). Accelerations with a uniform appearance occur simultaneously with uterine contractions, which provoke fetal movements (Figs 7.7 and 7.9); these are termed periodic accelerations.

Sporadic accelerations can also occur with the fetal movements provoked by obstetric procedures, such as manipulation of uterus and fetus, application of a scalp electrode, scalp blood sampling for pH, palpation of fetal parts at vaginal examination, and by other forms of external stimulation, such as vibroacoustic noise (Figs 7.10, 7.11 and 7.12); sporadic accelerations can also be spontaneous. Periodic accelerations with a rounded configuration (Fig. 7.13) were initially

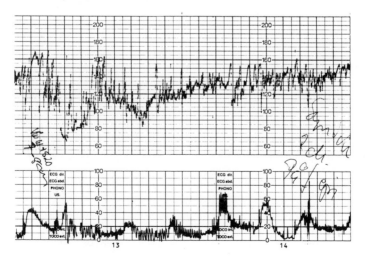

Fig. 7.4 Frequent accelerations and high FHR variability.

Accelerations

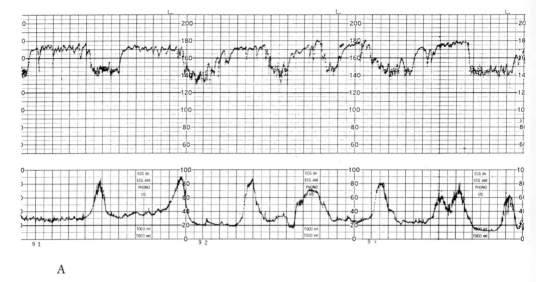

A

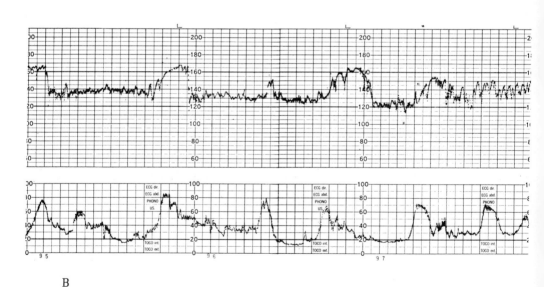

B

Fig. 7.5 A, Transient episodes of tachycardia (between 150 and 170 bpm) associated with fetal activity during labour. Misinterpretation of the record as late decelerations would result in unnecessary intervention; B, continuation of the same record showed that the true baseline was in the normal range. Problems like this are usually resolved if the record is prolonged.

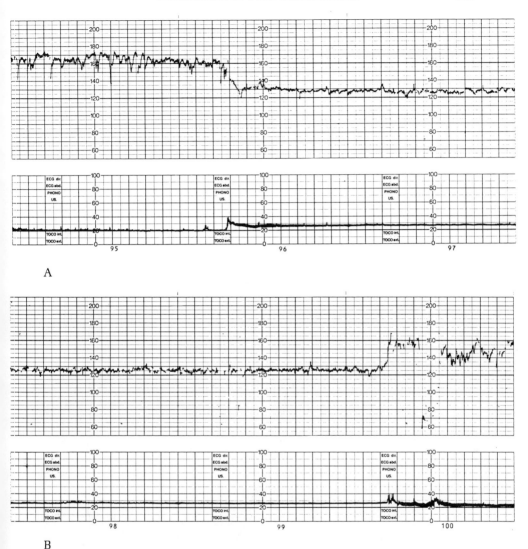

Fig. 7.6 A, A transient tachycardia, associated with frequent fetal movements (active fetal state) seen on the toco trace, was followed by an abrupt change to a quiet state lasting 40 min; B, after the quiet episode there was an abrupt resumption of accelerations and high variability characteristic of fetal activity.

considered to be indicators of fetal compromise because they are not common in labour.

Accelerations are a sign of a healthy autonomic nervous system and it has been shown in animal experiments and in human labour that accelerations are

Accelerations

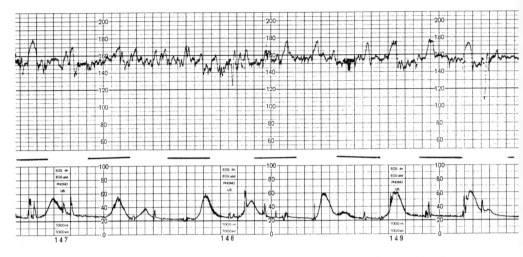

Fig. 7.7 Sporadic accelerations in labour, associated with evidence of fetal activity of the toco trace between the contractions.

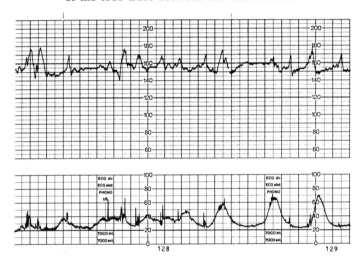

Fig. 7.8 Sporadic accelerations of variable appearance associated with fetal movements during labour.

absent if the fetus is acidotic. An acceleration may be interrupted by a variable deceleration secondary to compression of the umbilical cord during contractions. It is not uncommon, in cases with cord complications, to find a pattern of periodic accelerations in early labour, which later change to variable decelerations. The periodic accelerations then form a part of the typical variable deceleration (Fig. 7.14) and indicate well-being in a fetus adapting successfully to the transient episodes of hypoxia during cord compression.

Mechanism of FHR accelerations

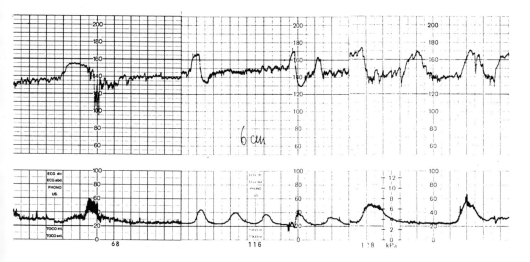

Fig. 7.9 Periodic accelerations with a more uniform appearance associated with contractions during labour.

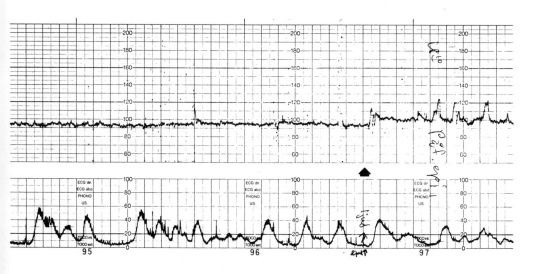

Fig. 7.10 A non-reactive CTG during labour showing reduced variability and a baseline FHR of 95 bpm. The pattern was observed for 1 h. Vibroacoustic stimulation (at the arrow) produced an acceleration and thereafter a reactive pattern was seen.

Accelerations

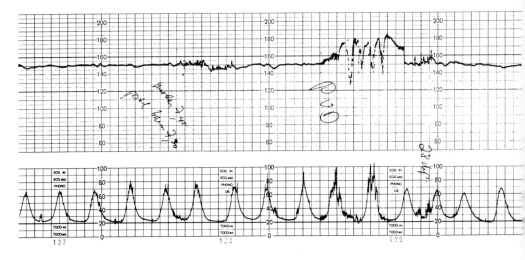

Fig. 7.11 A CTG with very low variability (which lasted several hours) showing the absence of an acceleration with a fetal blood sample. However, the pH was normal (7.30) and 10 min later, at vaginal examination for a pudendal block (marked PUD), accelerations were seen.

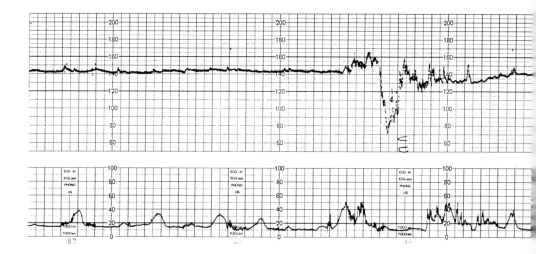

Fig. 7.12 A CTG with low variability followed by a combined acceleration and deceleration during a vaginal examination (marked VU). Thereafter, a reactive FHR was seen.

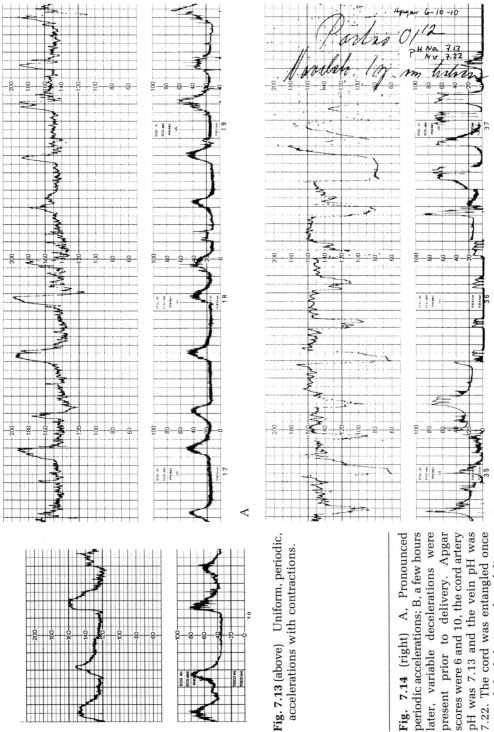

Fig. 7.13 (above) Uniform, periodic, accelerations with contractions.

Fig. 7.14 (right) A, Pronounced periodic accelerations; B, a few hours later, variable decelerations were present prior to delivery. Apgar scores were 6 and 10, the cord artery pH was 7.13 and the vein pH was 7.22. The cord was entangled once around the baby's neck at delivery.

Clinical importance of accelerations

In antepartum CTG records the presence of accelerations, associated with perceived fetal movements, is the basis for the description of the pattern as reactive. In some scoring systems, periodic accelerations were given a less favourable score than sporadic accelerations. However, such a distinction has not been shown to be of clinical importance and most scoring systems do not differentiate between the two types of accelerations. Scoring systems have never been popular in the UK and the term 'reactive' does not distinguish between types of accelerations.

The importance of accelerations of the FHR in labour has received insufficient attention. The occurrence of accelerations is an important indication that the fetus is not acidaemic, even if other abnormal signs are present (Fig. 7.15). Similarly, if three sporadic accelerations are present during the last 30 min of monitored labour the risk of the baby having a low 5' Apgar score is also very small. Accelerations provide reassurance when present with decelerations, especially if normal variability is also seen (Fig. 7.16). Absence of accelerations for more than 1 h during labour should serve as a warning, even if the trace is otherwise normal (assuming the fetus is not depressed by drugs). Accelerations and FHR variability are closely related but it is possible that the disappearance of accelerations is an earlier sign of fetal hypoxaemia than reduced FHR variability.

FHR variability is usually normal (between 10 and 25 bpm) when accelerations occur. However, traces can occasionally be seen with normal variability and

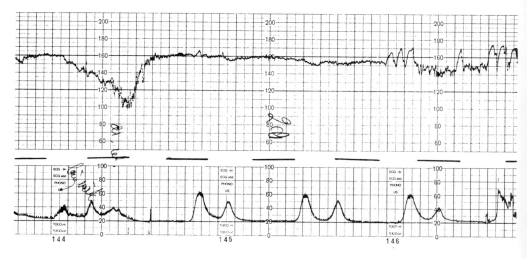

Fig. 7.15 An episode of mild tachycardia and low variability with one prolonged deceleration which appears worrying. Shortly after this the trace became normal and reactive.

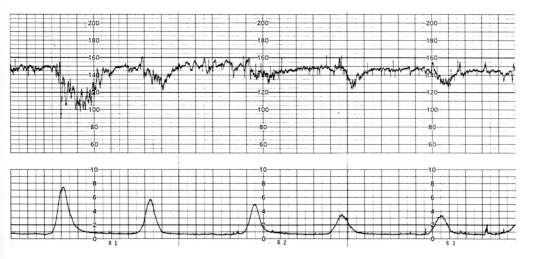

Fig. 7.16 Repetitive decelerations related to strong uterine contractions (intra-uterine pressure). However, reassurance was provided by the normal variability and the presence of accelerations.

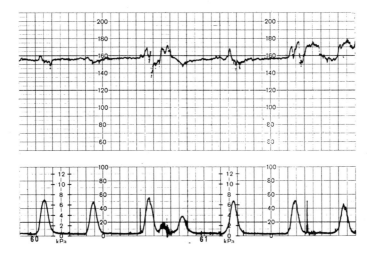

Fig. 7.17 Periodic accelerations with contractions despite very low variability.

no accelerations. It is more uncommon to find the presence of accelerations in traces with reduced or absent variability (Fig. 7.17). As a rule, traces containing accelerations, particularly sporadic ones, should not cause any concern.

Summary

The presence of accelerations indicates fetal well-being. However, the absence of accelerations is not a reliable indicator of fetal hypoxaemia and it is appropriate to continue the CTG recording in such cases.

8
Uniform decelerations

Introduction

A deceleration is a short-lasting reduction in FHR, usually falling more than 10 to 15 bpm, lasting for more than 15 s but less than 2 min. During pregnancy, decelerations are usually of pathological significance, whereas in labour they are quite commonly associated with uterine contractions.

Classifications

A number of classifications have been described for FHR patterns with decelerations, and the terminology is confusing. The most commonly used descriptive system is that of Hon (1968), which will be used in this book. Caldeyro-Barcia and co-workers' description (Caldeyro-Barcia et al. 1966) is also widely known: their 'type I dips' correspond to Hon's 'early' (uniform) or variable decelerations, or a combination of both, and their 'type II dips'

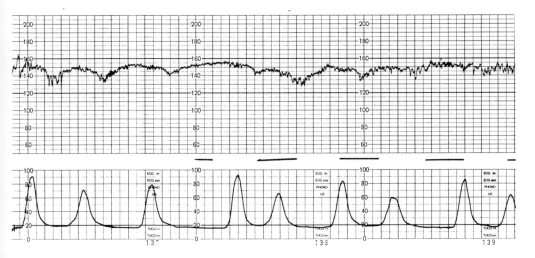

Fig. 8.1 A CTG with scalp electrode FHR and intrauterine pressure. The baseline FHR is 155 bpm with no accelerations, low variability and late decelerations.

Uniform decelerations

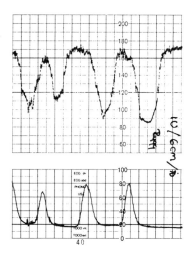

Fig. 8.2 The same case as shown in Figure 8.1 but 3 h later. The baseline FHR is now 170 bpm, still with low variability. The late decelerations are now pronounced. Decelerations like this may also be preceded by mild to moderate variable decelerations.

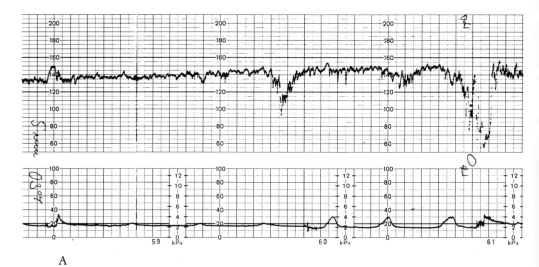

A

Fig. 8.3 The CTG of a multiparous woman at 39 weeks gestation whose labour was induced by artificial rupture of membranes (ARM) and oxytocin infusion because of a suspicious antenatal CTG and suspected fetal growth retardation. A, Scalp electrode FHR and external tocography after 20 min of oxytocin infusion (5 mU/min). The baseline FHR is 140 bpm with normal variability and late decelerations; B, 10 min later the FHR is between 130 and 140 bpm with normal variability and pronounced late decelerations. Fetal scalp blood pH was 7.31 and

(continued)

Classifications

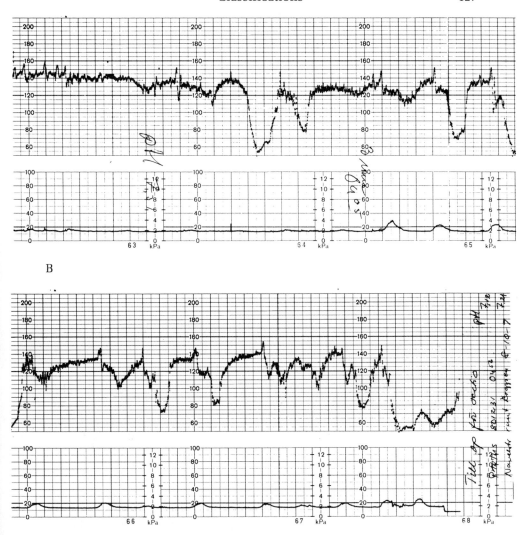

Fig. 8.3 *(continued)* the oxytocin was recommenced at 3 mU/min; C, continuation of the trace from B showing pronounced decelerations of combined (early and late) type. A prolonged deceleration (bradycardia) can be seen at the end of the record. A healthy baby was delivered at emergency Caesarean section. Apgar scores were 8 and 9, the umbilical artery pH was 7.12, the vein pH was 7.21 and birthweight was 2350 g. This case illustrates successful fetal adaptation to a chronic reduction in placental function (small-for-gestational age and probably growth-retarded) until acute hypoxia was superimposed during uterine tightenings. The normal rate and variability, with some reactivity, indicate a well oxygenated fetus, this was confirmed at birth.

correspond to Hon's late (uniform) decelerations. Both of these descriptive classifications imply that the temporal relationship between deceleration and uterine contraction is important.

Others have questioned the relevance of this relationship, claiming that the number of heart beats lost (dip area) during a deceleration is more important. According to this proposal, all fetuses will develop distress after a sufficient number of lost beats. However, a major flaw with this approach is that shallow, late, decelerations, which are particularly associated with severe, chronic, fetal hypoxia during pregnancy, would not be regarded as sinister. This pattern is found in about 5 per cent of CTG records obtained between 28 and 32 weeks and occurs, classically, with severe pregnancy-induced hypertension associated with intrauterine growth retardation (Figs 8.1, 8.2 and 8.3). It is less useful to distinguish between uniform or variable late decelerations, especially when pronounced, as such patterns are equally associated with fetal distress (Fig. 8.2).

Characteristics

Decelerations are described according to their shape because, on the whole, there is a reasonable relationship between shape and causation. The shape of the deceleration usually allows use of the major descriptive terms 'uniform', 'variable' (see Chapter 9), or 'combined' (see Chapter 10). The temporal relationship to uterine contractions is of relevance if the decelerations are uniform.

Uniform decelerations usually have a slow onset and recovery, which gives them a rounded configuration, classically appearing as an inverted image of the shape of the uterine contraction. Accelerations at the beginning or end of such decelerations are uncommon. The size (depth) of the deceleration is often well correlated with the amplitude of the contraction and, as a rule, deeper decelerations occur with stronger contractions. The uniform shape is, however, the same even if the amplitude of the contraction varies (Fig. 8.4).

The shape of variable decelerations varies from one contraction to the other (see Fig. 9.1). The abrupt fall in FHR and an equally fast recovery suggests a nervous system reflex and the parasympathetic nervous system (vagus nerve) is the likely effector. Changes of maternal and fetal position can influence the shape of variable decelerations, which suggests that their origin relates to placental perfusion, either compression of the umbilical cord vessels or interference with uteroplacental perfusion. An acceleration at the onset, as well as after these decelerations, is associated with fetal well-being and is likely to reflect interruption of a contraction-related acceleration by cord compression (Fig. 8.5). The temporal association of variable decelerations to uterine

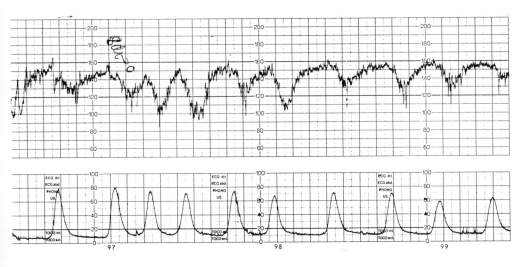

A

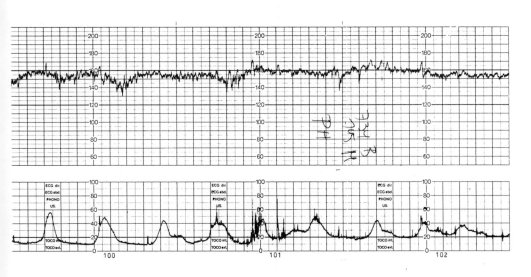

B

Fig. 8.4 The fetal ECG and intrauterine pressure record from a primipara at 38 weeks gestation whose labour was augmented with oxytocin. A, The oxytocin infusion was stopped (at mark 97) because of late decelerations. The baseline FHR was between 150 and 155 bpm with normal variability; B, the late decelerations disappeared as the amplitude and frequency of uterine contractions decreased. Fetal scalp blood pH was normal (7.34).

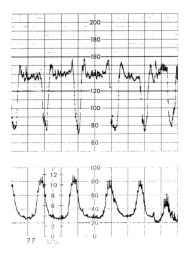

Fig. 8.5 Typical short, early, variable decelerations.

contractions may be early or late but this is not a useful distinction because, in some cases, it is variable. Sometimes, the onset of a variable deceleration is related to the onset of contraction, sometimes at the peak, and sometimes the end. Early decelerations, with abrupt FHR changes of large depth, are not common in the first stage of labour. Early uniform and variable decelerations may be seen in the same record (Fig. 8.6).

The timing of uniform decelerations

The time relation between deceleration and contraction is of major importance with regard to the classification of uniform decelerations into early and late. An early uniform deceleration has an onset before the peak of the contraction, whereas a late uniform deceleration starts at the peak or later. This is easy to see while standing by the monitor watching the continuous printout (Fig. 8.7).

Early uniform decelerations

This type of deceleration begins shortly after the onset of the contraction but before its peak. The origin is believed to be pressure from surrounding tissues (uterus, pelvic floor, perineum) on the fetal head during the contraction. Pressure on the fetal head raises intracranial pressure and may affect cerebral blood flow. Vagal tone rises as the pressure rises and this produces a slow-onset,

The timing of uniform decelerations

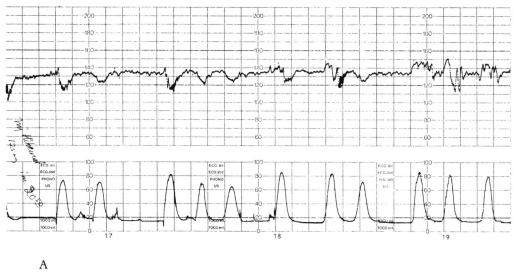

A

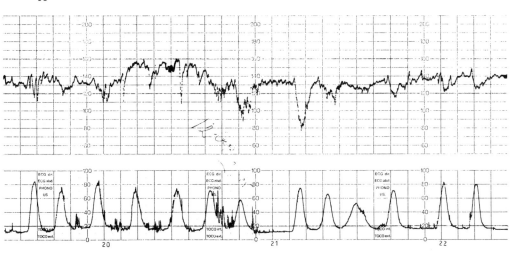

B

Fig. 8.6 A fetal ECG and intrauterine pressure record. Both early and variable decelerations can be seen in A and B.

reflex-mediated fall in FHR. As the peak of the contraction is reached, so the deceleration becomes maximal and, thereafter, the FHR returns to normal as the contraction and associated head pressure eases off (Fig. 8.8).

Early decelerations often appear within the normal baseline FHR (between 110 and 150 bpm) and are not usually pronounced during the first stage of labour

Uniform decelerations

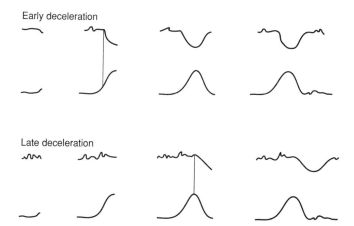

Fig. 8.7 A schematic figure showing the relationship between deceleration and uterine contraction. The vertical line represents the start of the deceleration. Early decelerations begin with the onset of the uterine contraction and usually have their trough with the contraction peak. Late decelerations begin after the onset of the contraction and have their trough after the contraction peak.

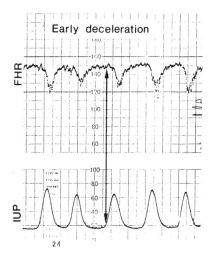

Fig. 8.8 An FHR and intrauterine pressure record showing early decelerations.

(Fig. 8.9). Such decelerations seldom fall below a rate of 100 bpm. The duration is rarely longer than 60–90 s and never longer than the uterine contraction. They are innocuous and seldom associated with fetal hypoxaemia or acidosis (Fig. 8.10). Early decelerations appear more frequently after rupture of membranes.

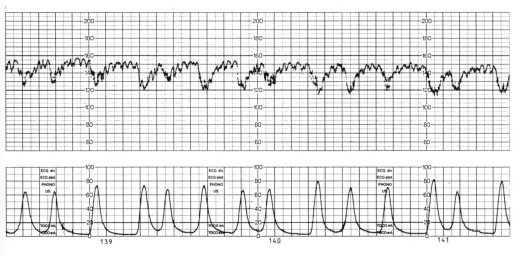

Fig. 8.9 Typical early decelerations.

Late uniform decelerations

Late uniform decelerations are of more clinical importance than early uniform decelerations. The onset of late decelerations is usually after the onset of the contraction and may even be after the peak of the contraction. The nadir of a late uniform deceleration is more than 15 s after the peak of the contraction and so recovery from the deceleration is not usually complete until well after the uterus has relaxed (Figs 8.7 and 8.11). The decelerations are of similar (uniform) appearance from one contraction to another but may have a more rounded shape than early uniform decelerations. A 'hammock' appearance may be seen when late uniform decelerations are pronounced (Fig. 8.12). The depth of the deceleration represents the degree of fetal response to the acute component, which occurs during each contraction.

Late decelerations may also be classified, according to lost beats, into three groups: less than 15 bpm, between 15 and 45 bpm, and greater than 45 bpm. Late decelerations usually have a moderate loss of beats and such decelerations seldom fall below a rate of 100 bpm. Pronounced late decelerations may fall as much as 50–60 bpm but the duration is often less than 90 s. Late decelerations quite often start from a raised baseline FHR of between 150 and 170 bpm (Figs 8.11 and 8.12) and are certainly an occasion to consider a baseline FHR above 150 bpm as pathological.

The origin of late decelerations is not completely understood, although much information has been derived from animal experiments. It is generally agreed that late decelerations occur as a result of contractions when the fetus is already significantly hypoxic (acute-on-chronic fetal hypoxia). The late deceleration is

Uniform decelerations

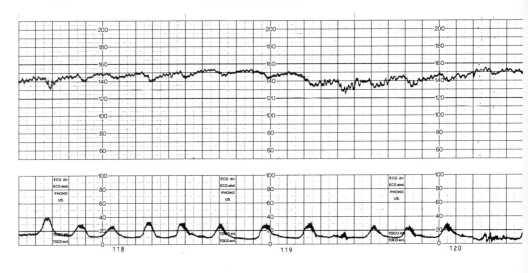

A

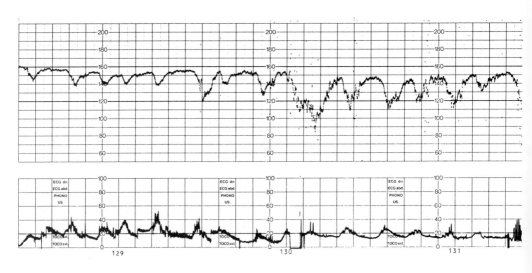

B

Fig. 8.10 Spontaneous labour in a multipara at 41 weeks gestation. A, At cervical dilatation 7 cm, the baseline FHR was 150 bpm with low variability, no accelerations and early decelerations. Fetal scalp blood pH was 7.32; B, after a further 80 min, the cervix was 9 cm dilated with the vertex at the ischial spines. Baseline FHR was between 150 and 155 bpm with low variability and early and late decelerations.
Delivery occurred 20 min after this trace. Apgar scores were both 10.

The timing of uniform decelerations

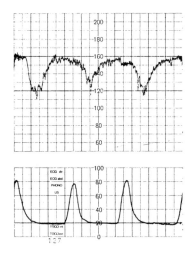

Fig. 8.11 Typical late decelerations.

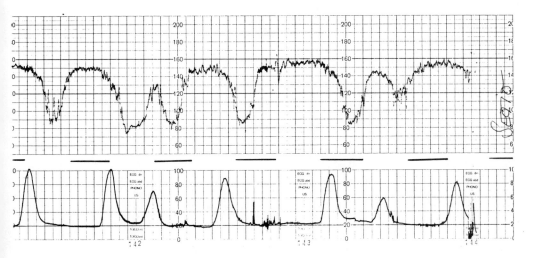

Fig. 8.12 Pronounced late decelerations.

modified, but not abolished (as with early decelerations), by a parasympathetic blocking agent, such as atropine, which indicates that the vagus nerve is not the only contributing mechanism in these circumstances. Hypoxic depression of the myocardium is considered the most likely mechanism, leading to reduced cardiac output and reduced capacity of the fetus to respond to the hypoxic situation. A good correlation exists in animal experiments between late decelerations and the arterial oxygen tension of the fetal blood but not with pH. Thus, the obstetrician cannot expect a straightforward correlation between

136 Uniform decelerations

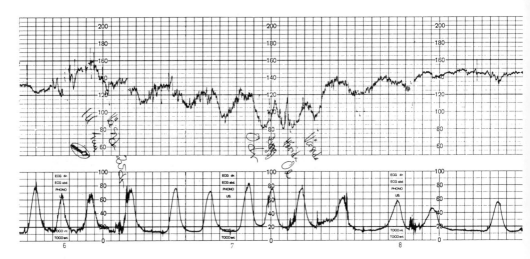

Fig. 8.13 The FHR record in a primipara following induction of labour by ARM and oxytocin infusion. At the beginning of the recording a vaginal examination provoked a pronounced FHR acceleration, suggesting well-being. The cervix was dilated to 4 cm. Associated with lying supine, late decelerations appeared and the FHR fell over the next 10 min. The oxytocin infusion was stopped at this point and the FHR returned to normal. The late decelerations disappeared. Fetal scalp blood pH was 7.3. A normal spontaneous delivery occurred 5 h later; the umbilical artery pH was 7.21 and the vein was pH 7.31. Apgar scores were 9 and 10.

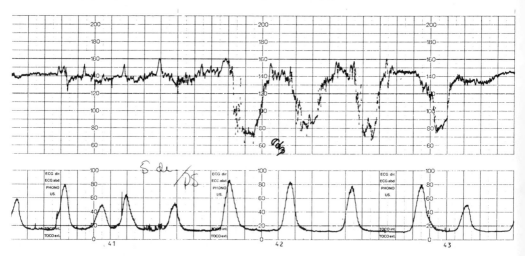

Fig. 8.14 A primipara in slow labour at term. Intrauterine pressure was recorded. Oxytocin (5 mU/min) was commenced at mark 41. The amplitude of the contractions increased immediately from between 40 and 50 mmHg to 70 mmHg. Pronounced late decelerations occurred with these but the baseline remained normal. The infusion was stopped at mark 42 and the FHR pattern returned to normal after this.

fetal blood pH and late decelerations. Clinical studies have shown that late decelerations are associated with fetal acidaemia in approximately 40 per cent of cases.

The cause for late decelerations is often clinically apparent but such decelerations may be present without other abnormal features on the tracing. Such circumstances are abnormal uterine activity, usually induced (Fig. 8.13), maternal hypotension (vena cava syndrome), or after epidural block. A normoxic fetus is capable of reacting to these situations with late decelerations (Fig. 8.14) and so pH sampling is required to distinguish a pathological from a physiological response. Late decelerations are usually associated with good baseline FHR variability and a normal FHR if the fetus is not acidaemic. Late decelerations will usually disappear by correcting the cause (stopping the oxytocin infusion, correcting the maternal supine position, restoration of maternal blood pressure, and increasing circulating volume).

During pregnancy, late decelerations are associated with chronic fetal hypoxia secondary to 'uteroplacental insufficiency', which results from a chronic reduction in the maternal placental circulation (acute pre-eclampsia, intrauterine growth retardation, and recurrent abruptio placentae). Late decelerations may appear in a pathological, non-reactive record with tachycardia and low baseline variability. It is likely that such late decelerations in complicated pregnancies, in the absence of Braxton Hicks tightenings, reflect hypoxic depression of fetal myocardial function. These decelerations, associated with abnormal placental function, are not likely to be altered by simple corrective measures and the fetal scalp blood pH is likely to be below normal if such cases enter labour (Fig. 8.15).

Clinical importance of late decelerations

The interpretation of late decelerations in labour is difficult as they may be the result of one of a number of mechanisms that interfere with placental function. The clinical circumstances are important and must be evaluated before a decision about the significance of the CTG is reached. Even then, it may still be necessary to perform fetal blood sampling to identify late decelerations resulting from obstetric manoeuvres. Excessive intervention has been shown to result from policies of continuous monitoring in labour without fetal blood sampling. Figure 8.16 shows such an example of late decelerations due to abnormal uterine activity induced by oxytocin infusion. A non-reactive trace with low baseline variability and tachycardia suggests fetal stress. The risk for fetal acidosis is significant if the sample is taken before reduction of the oxytocin infusion. Such iatrogenic fetal hypoxia should be corrected by reduction of the oxytocin infusion rather than obstetric intervention. Similarly, to intervene on the presumption of fetal hypoxia because of repetitive late decelerations for some time may be wrong. In most situations there is time to make a clinical evaluation of the patient.

Uniform decelerations

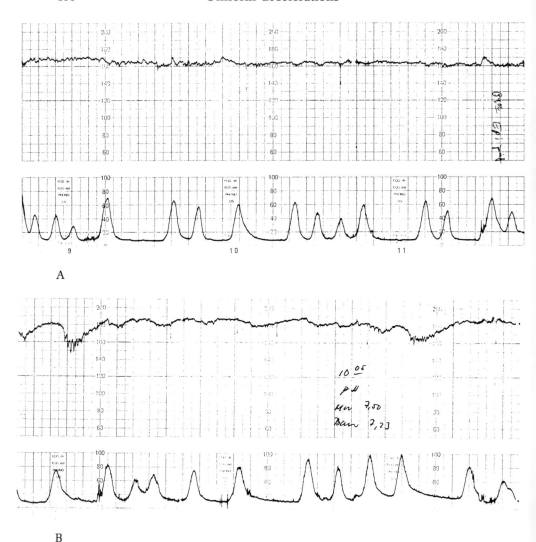

Fig. 8.15 (caption opposite).

All possible factors likely to produce late decelerations should be considered: infusion of oxytocin, use of prostaglandins, vena cava compression, conduction blocks for pain relief (spinal, epidural, paracervical), anaemia (maternal, fetal), pre-eclampsia, intrauterine growth retardation, vaginal bleeding (abruptio placentae). Associated factors may increase the chance of acidaemia being present: prematurity, prolonged labour, oligohydramnios, meconium-stained liquor, fetomaternal disproportion, fetal malformation, fetal infection, and so on.

The timing of uniform decelerations

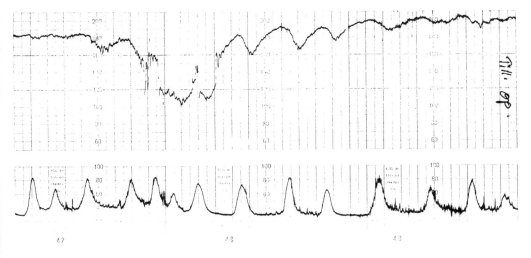

C

Fig. 8.15 A primipara at 40 weeks in spontaneous labour (intrauterine pressure record). A, After 5 h of normal CTG, a non-reactive tachycardia (165 bpm) developed with low variability but no decelerations. The fetal scalp blood pH was 7.39; B, a further 5 h later, a tachycardia (180 bpm) with late decelerations is evident. Variability is very low. The fetal scalp blood pH was 7.23; C, this record continues from B. After a prolonged deceleration the FHR rose further to reach 200 bpm. Late decelerations and very low variability persisted. An emergency Caesarean section was performed. The umbilical artery pH was 7.12 and Apgar scores were 5 and 9.

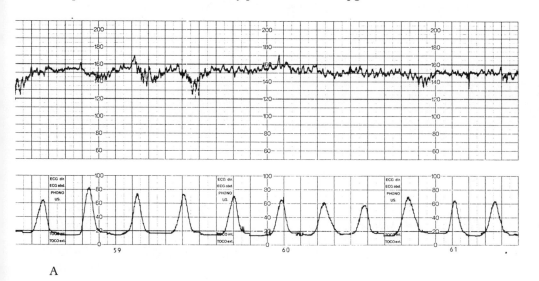

A

Fig. 8.16 (caption overleaf).

Uniform decelerations

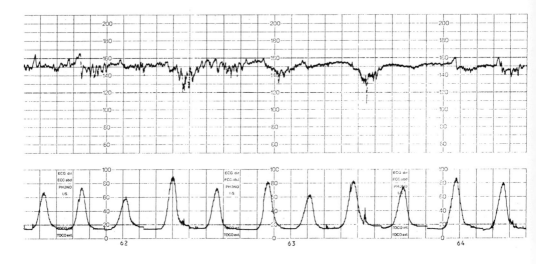

B

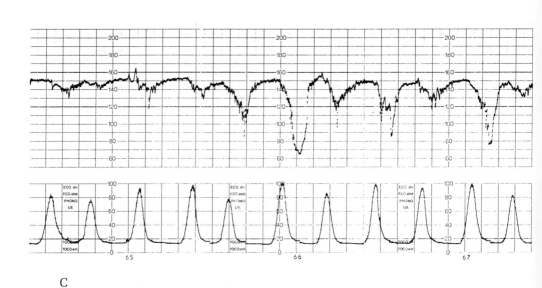

C

Fig. 8.16 Late decelerations induced by oxytocin. A, A normal FHR and uterine activity (intrauterine pressure measurements of contractions 60 mmHg); B, increased amplitude of contractions to between 70 and 80 mmHg. Late decelerations can just be seen; C, the amplitude of the contractions rose to between 80 and 100 mmHg. The baseline FHR remained unchanged but variability was reduced and the late decelerations increased in size.

The timing of uniform decelerations

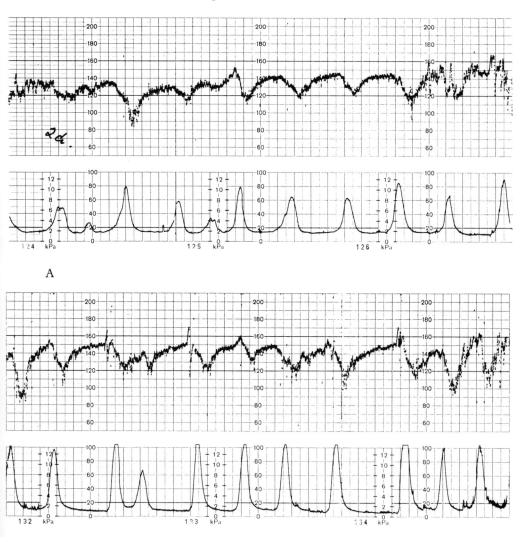

Fig. 8.17 (caption overleaf).

The management of cases with late decelerations depends, to a large extent, on the presence or absence of baseline variability and accelerations. The risk for fetal acidaemia is relatively small if these aspects are normal (Figs 8.17, 8.18 and 8.19). The situation is more serious in the presence of an FHR tachycardia and absence of variability or accelerations. In these circumstances it may be appropriate to consider intervention, particularly if it is not possible to obtain

Uniform decelerations

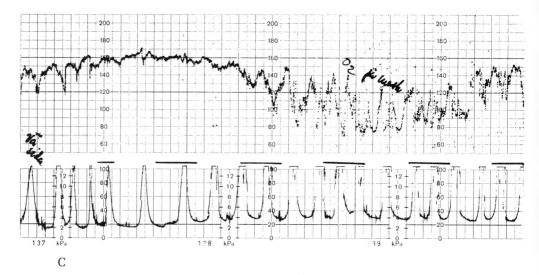

C

Fig. 8.17 A primipara at 40 weeks in spontaneous labour with ruptured membranes. Fetal ECG and intrauterine pressure record. A, An infusion of oxytocin was started because of slow progress at 5 cm. Repetitive late decelerations were seen; B, 60 min later, the fetal scalp blood pH was normal (7.34). Uterine hyperstimulation, with a mean amplitude of contractions about 100 mmHg, is evident and is accompanied by late decelerations; C, the last 30 min (20 min after the record in B) showed early decelerations in the second stage. The umbilical artery pH was 7.19 and Apgar scores were 9 and 10. Another example of hyperstimulation by oxytocin without the fetus becoming significantly acidaemic.

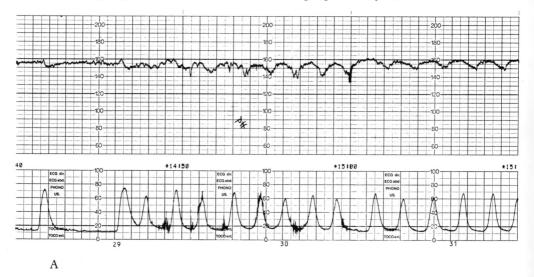

A

Fig. 8.18 (caption opposite).

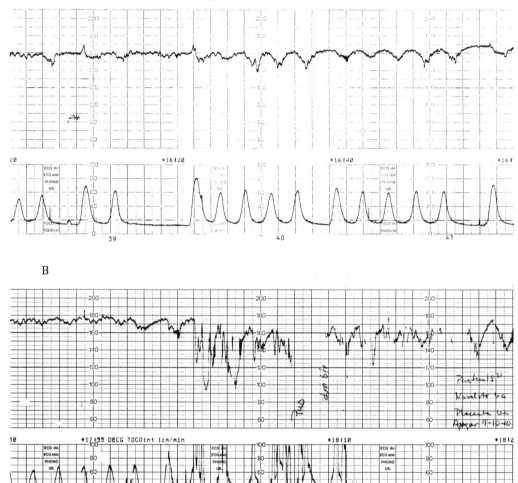

Fig. 8.18. A primipara at 40 weeks in prolonged labour with an oxytocin infusion and a mild tachycardia at panel A (cervix dilated 8 cm). A, The fetal head was at the spines. The rate of oxytocin infusion was 20 mU/min. The intrauterine pressure record showed frequent contractions, with a mean intensity of between 60 and 70 mmHg. The FHR variability was low with recurrent late decelerations but fetal scalp blood pH was 7.3 (at mark 30); B, 1.5 h later. The variability was still low and repetitive late decelerations persisted. Fetal scalp blood pH was 7.28 (at mark 39); C, 1 h and 15 min later, prior to delivery. Early and late decelerations were seen during pushing. Apgar scores were 9 and 10. This case illustrates the need for fetal blood sampling to exclude fetal acidosis.

Uniform decelerations

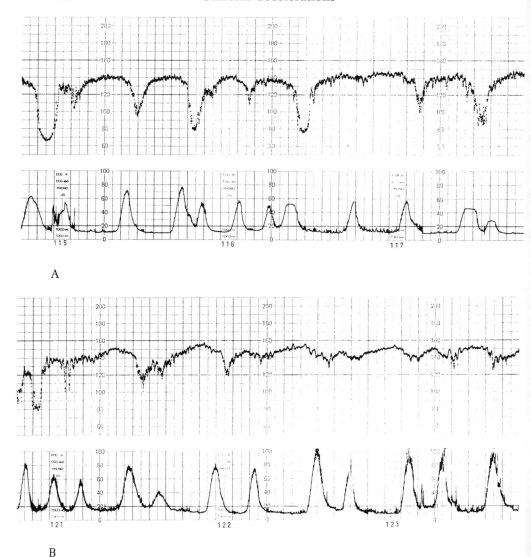

Fig. 8.19 (caption opposite).

a fetal blood sample for pH (Figs 8.20 and 8.21). FHR variability during the deceleration is also of importance and absence of variability is more often associated with fetal acidaemia (Figs 8.22, 8.23 and 8.24).

Progressive fetal hypoxia may be associated with a CTG such as in Figure 8.25. In contrast, late decelerations may suddenly appear in an otherwise normal CTG (Fig. 8.26) with acute fetal hypoxia (such as abruptio placentae). Late

The timing of uniform decelerations

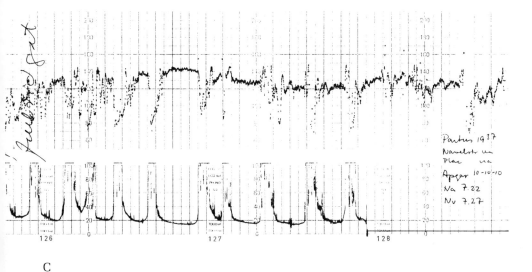

C

Fig. 8.19 The CTG of a primipara at 38 weeks in spontaneous labour. The FHR was obtained from the ECG and uterine acivity was monitored by intrauterine pressure. The cervix was dilated 6 cm and the fetal head was just below the level of the spines. A, Normal FHR but low variability and late decelerations. Fetal scalp blood sampling just after this trace was normal (pH 7.31). Not all contractions were associated with late decelerations. The intrauterine catheter become blocked at the end of the record detected by a 'non-physiological' contraction shape; B, 40 min later the cervix was fully dilated. The FHR was unchanged with less pronounced late decelerations despite a similar uterine activity; C, the second stage with early decelerations prior to delivery. The umbilical artery pH was 7.22 and the vein pH was 7.27. Apgar scores were normal. Improvement of the FHR pattern during the progress of labour meant that repeated fetal blood sampling was not necessary.

decelerations may be managed without intervention if the risk for acidaemia is relatively low and the alternative course of action impractical, e.g. a multiparous woman in late first stage of labour if vaginal delivery is likely to be quicker than a Caesarean section. The presence of hypoxia does not automatically imply that the fetus will be acidaemic; delivery may occur before acidaemia develops.

The situation is different for a primiparous woman in prolonged labour, particularly in the second stage with an increased risk for fetal hypoxia. It may be appropriate to consider Caesarean section delivery, even if the fetal blood pH is normal but the values are falling and are likely to reach abnormal levels before delivery. Figure 8.27 shows such a case with late decelerations, low baseline variability, and a tachycardia but a normal scalp blood pH (7.3). This primiparous woman had slow progress of labour, despite augmentation with oxytocin, and the cervix dilated to only 4 cm. A Caesarean was performed

146 Uniform decelerations

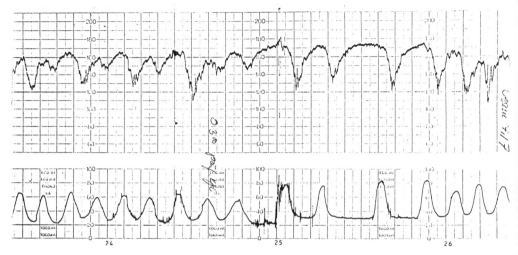

Fig. 8.20 FHR (by ECG) and extrauterine pressure. The tachycardia (FHR between 165 and 170 bpm), very low variability and repetitive late decelerations indicate an increased risk (between 40 and 50 per cent) of fetal acidosis. This was confirmed by fetal scalp blood sampling (pH 7.17).

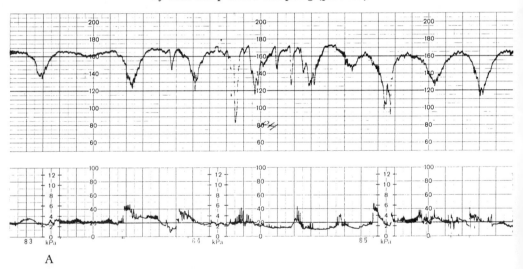

A

Fig. 8.21 A primipara at term in spontaneous labour. A, Cervix dilated 5 cm and the fetal head just above the level of the spines. Uterine activity was monitored by external tocodynamometry. The CTG showed a tachycardia (between 165 and 170 bpm), very low variability and repetitive late decelerations. Fetal scalp blood pH (at mark 64) was 7.24; B, 40 min later. The cervix was 9 cm dilated and the fetal head was at the spines. A pronounced tachycardia (between 170 and 180 bpm), absent variability and repetitive late decelerations was seen. The fetal scalp blood
(continued)

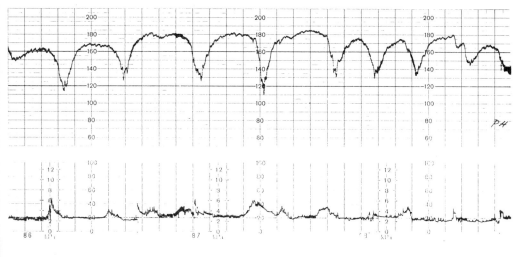

B

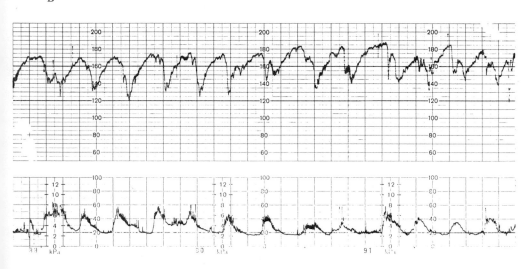

C

Fig. 8.21 (continued)   pH was 7.16. A trial of vaginal delivery was planned due to a severe maternal complication; C, this trace continues from B and shows slightly improved variability. A forceps delivery was performed 20 min after this trace. The umbilical artery pH was 6.96. Apgar scores were 3, 4, and 5 at 1, 5, and 10 min, respectively. The long-term outcome for the infant was normal. This is an example of a typical pathological FHR pattern with tachycardia, absence of variability and late decelerations. Although there was a high risk of fetal asphyxia, with fetal acidosis and depression at birth, a favourable long-term outcome indicates this was not severe or protracted.

148 Uniform decelerations

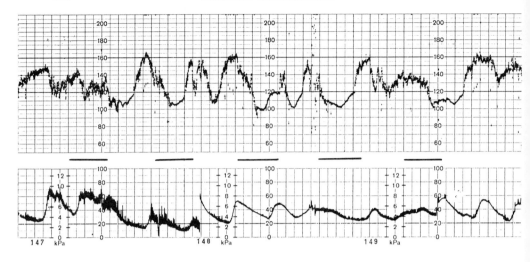

Fig. 8.22 Prolonged late decelerations with reduced variability during the decelerations.

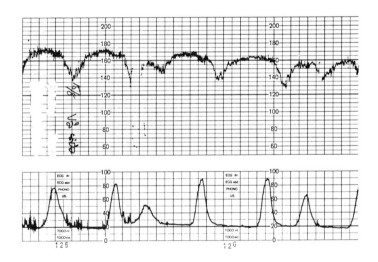

Fig. 8.23. Late decelerations with normal variability during the decelerations.

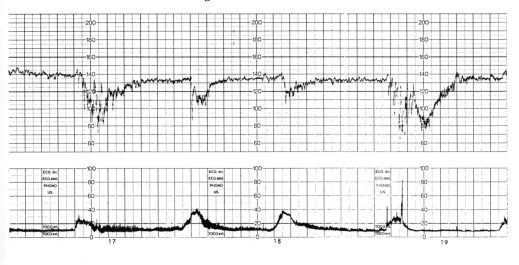

A

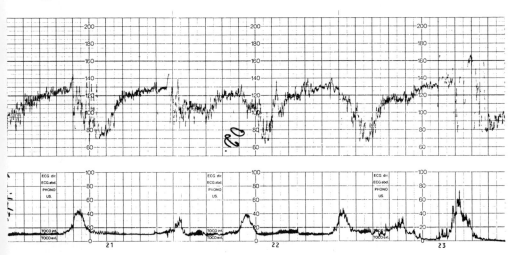

B

Fig. 8.24 A primipara at term in spontaneous labour after rupture of the membranes. Uterine activity was monitored externally. A, A normal FHR and low variability between contractions. The variability increased during the late decelerations. Fetal scalp blood pH was normal (7.30); B, 10 min later, longer duration late decelerations occurred. 10 min after this panel, fetal scalp blood pH had fallen to 7.20. An emergency Caesarean section was performed. Umbilical artery pH was 7.24 and venous pH was 7.31. Apgar scores were 9 and 10. The increased variability probably indicated acute hypoxia which may have been secondary to cord compression.

150 Uniform decelerations

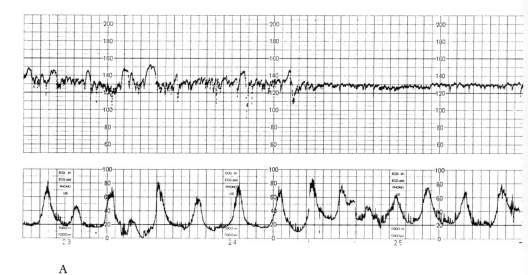

A

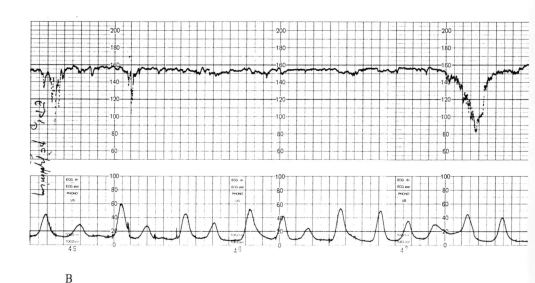

B

Fig. 8.25 (caption opposite).

because of abnormal progress and the assumption that the fetus was at high risk of developing hypoxia because of prolonged labour. Repetitive scalp blood pH determinations are appropriate if good progress towards a vaginal delivery continues.

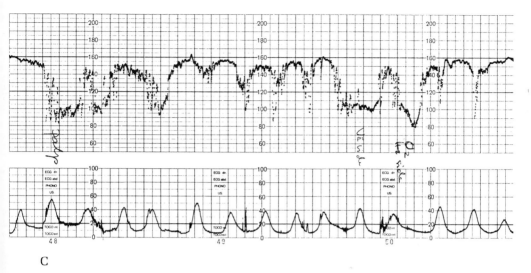

C

Fig. 8.25 Progressive fetal hypoxia. A, Early labour. The CTG clearly shows evidence of fetal behaviour, with a change from activity to quiescence; B, some while later, tachycardia and low variability are evident. There are no accelerations; C, further on, prolonged and late decelerations are an additional risk factor for the presence of fetal hypoxia and acidaemia. Repetitive determination that fetal blood pH remains normal will be necessary for conservative management.

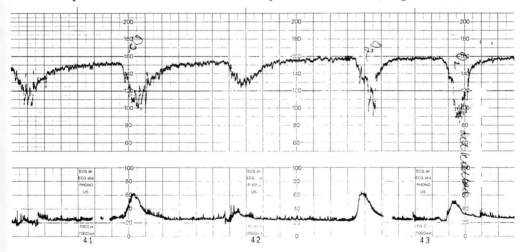

Fig. 8.26 A multipara with severe polyhydramnios at term in labour. Late decelerations occurred following the ARM and a prolonged deceleration. The FHR was 155 bpm and variability was low. Fetal scalp blood pH, soon after this tracing, was 7.24. A Caesarean delivery was performed because of vaginal bleeding. A suspected placental abruption was confirmed at operation. Apgar scores were 9 and 10. This illustrates late decelerations without fetal distress.

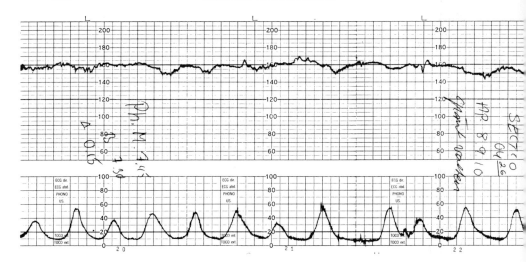

Fig. 8.27 Late decelerations, low baseline variability, tachycardia and normal scalp pH (7.3).

Summary

The normal fetus has considerable resources and may compensate for moderate hypoxia, indicated by FHR decelerations, for some time. Reassurance of fetal well-being during labour complicated by decelerations requires a normal FHR, variability, and the presence of FHR accelerations, without which such reassurance must be ascertained by the rate of change of fetal scalp pH.

9
Variable decelerations

Introduction

The variable deceleration is the most common type of FHR deceleration, appearing in 25–30 per cent of all labours. It causes considerable anxiety in clinical practice. Variable decelerations indicate a fetal response to acute hypoxia (usually related to a uterine contraction) but do not indicate the overall fetal oxygen level. If placental function is normal then the fetus may remain well oxygenated despite large or recurrent variable decelerations. If the degree of hypoxia during each contraction exceeds the duration or ability of the placenta to achieve full recovery of the oxygen level between contractions, then sustained fetal hypoxia may develop. The speed of development of fetal hypoxia will relate to placental function as well as to the specific cause of the hypoxia. In contrast to late decelerations, the outcome of variable decelerations is hard to predict. As their description implies, they are variable both in occurrence and degree.

Aetiology

Variable decelerations are recognized by the variation in appearance from one contraction to another. They also vary in duration, timing, and intensity relative to each contraction. They indicate a fetal reflex response by the nature of the abrupt fall in heart rate, often down to between 60 and 80 bpm, and the subsequent rapid return to baseline. They are often preceded and followed by a rise in FHR, which gives the appearance of an interrupted FHR acceleration (Fig. 9.1).

Variable decelerations are considered to be most commonly the result of compression of the umbilical cord and are sometimes described as cord dips; animal experiments and clinical observations support this. Clamping of the cord in animal experiments provokes an immediate and abrupt fall in FHR, similar in onset to that of a variable deceleration. Selective clamping of the umbilical vein produces an initial acceleration (Fig. 9.2).

It has been suggested that the cord compression begins with the thin-walled umbilical vein, which produces the initial FHR acceleration. However, it is possible that the FHR acceleration occurs as discussed in Chapter 7. Interruption of the return of oxygenated blood to the heart is likely to trigger the

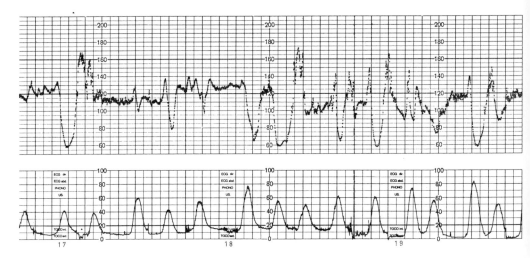

Fig. 9.1 Typical variable decelerations. The uterine activity record is intrauterine pressure. In contrast to uniform decelerations, variable decelerations are not related to the intensity of the contractions.

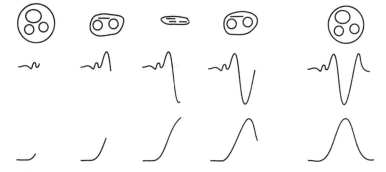

Fig. 9.2 A schematic picture of the FHR response to umbilical cord compression.

chemoreceptors and produce an abrupt vagal bradycardia. If the thicker-walled arteries become compressed, as the pressure increases during the contraction, this would trigger the baroreceptors, which register the increase in blood pressure secondary to the increased peripheral resistance. This probably augments the vagal reflex bradycardia. The rapid return to normal FHR, after release of the compression, further indicates the likelihood that the deceleration is a reflex response.

More information about variable decelerations has been obtained using Doppler ultrasound and analysing the fetal blood velocity waveform. End diastolic blood flow in the umbilical artery may be abolished, or even reversed,

Aetiology

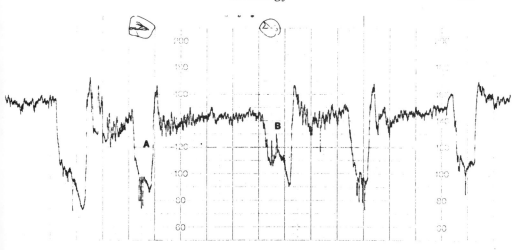

Fig. 9.3 Variable decelerations, probably due to umbilical cord compression (see text).

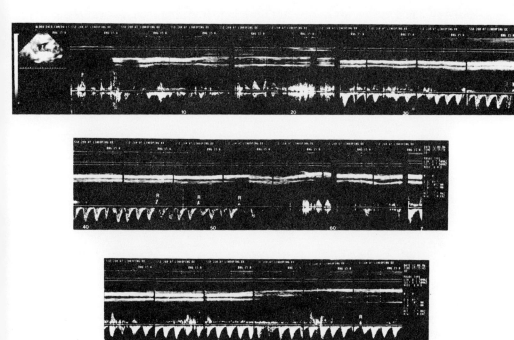

Fig. 9.4 Record of umbilical artery blood flow during a variable deceleration (marked B on the FHR record shown in Figure 9.3). Reversed diastolic flow is identified by R.

in cases of cord compression. Figure 9.3 shows variable decelerations in a case where a loop of the cord could be seen between the uterine wall and the fetal head. High resistance to blood flow was measured by Doppler ultrasound and a sector scanner, as indicated by reduced or absent end diastolic velocities, or even negative velocities (Fig. 9.4). The blood flow analysis in Figure 9.4 represents 45 s of the deceleration marked B in Figure 9.3.

It is often not possible to prove that cord compression has occurred in labours with variable decelerations. Conversely, many deliveries occur with the umbilical cord in several loops around the neck without there having been variable decelerations during labour. The length of the cord is important. Thirty per cent of labours with a short umbilical cord (< 35 cm), and more in cases with a long cord (> 80 cm), are associated with variable decelerations. The volume of amniotic fluid is also important, as the cord is more likely to be compressed if there is reduced amniotic fluid. In cases of intrauterine growth retardation, a reduction of the protective Wharton's jelly may also predispose the cord to compression.

As mentioned previously, variable decelerations may interrupt periodic accelerations that occur with contractions. Thus, the acceleration, indicating fetal well-being, begins at the onset of the contraction and then, as the pressure builds up and cord compression occurs, there is a sudden rapid fall in FHR. As the contraction pressure subsides, the FHR returns rapidly to the point at which it would have been had the acceleration not been interrupted (Fig. 9.5).

Variable decelerations may also appear if the fetal head descends rapidly through the birth canal. Such variable decelerations have a low amplitude and a short duration. Variable decelerations are also frequently seen during second

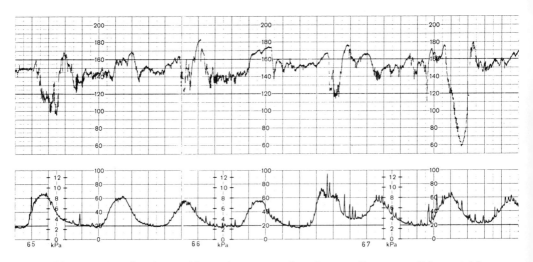

Fig. 9.5 Accelerations with uterine contractions become 'interrupted' by variable decelerations (paper speed 2 cm/min).

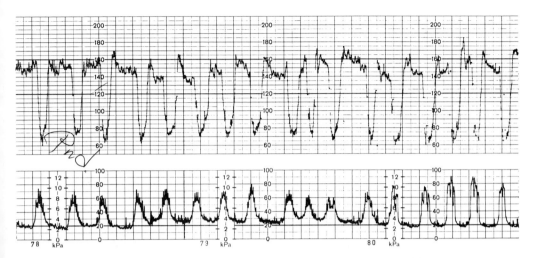

Fig. 9.6 Uterine activity monitored externally as full dilatation was reached. A pudendal block was given (marked 'Pud'). Moderate variable decelerations continued with a normal baseline and variability. The absence of ominous signs meant that fetal blood sampling was unnecessary.

stage of labour. Decelerations at this stage are often quite pronounced but have good FHR variability between them (Fig. 9.6).

Fetal oxygenation and acidaemia

Short periods of cord compression may be compensated in between contractions. In such cases, fetal oxygenation might not be affected at all despite variable decelerations concomitant with every contraction (Fig. 9.7). If acidaemia is developing, transient periods of tachycardia may follow each deceleration. The duration of such tachycardia varies between 30 and 90 s. The profile is often sloping, with decreased variability and a slow return to the baseline. This tachycardia, which may represent stimulation of catecholamine release from the adrenals, is quite different from an acceleration that may follow a variable deceleration (Figs 9.8 and 9.9). Occasionally, an 'overshoot' occurs after the deceleration which, if associated with loss of short-term variability, is more likely to be associated with fetal hypoxia (see Fig. 9.27).

The degree and duration of cord compression during contractions are important determinants of the development of fetal acidaemia. Mild cord compression will produce respiratory acidosis secondary to decreased CO_2 removal. This is not usually associated with neonatal depression. However, if prolonged and repetitive cord compression results in a significant reduction in

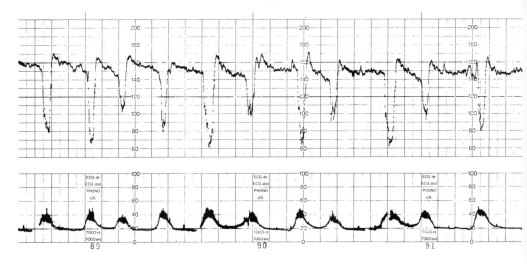

Fig. 9.7 Uterine activity monitored by external tocography. Moderate variable decelerations which caused no concern during the first stage.

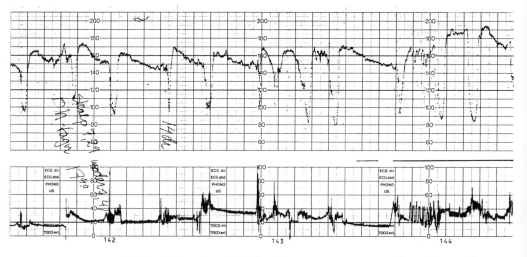

Fig. 9.8 Variable decelerations each followed by a compensatory tachycardia. Fetal scalp blood pH was 7.29. A spontaneous delivery occurred after 1 h and the baby was healthy.

fetal oxygen delivery then progressive fetal metabolic acidosis will develop. This situation is compounded if the interval between contractions is insufficient to allow adequate exchange of O_2 and CO_2 in the intervillous space. The associated redistribution of the fetal circulation in response to hypoxia results in hypoperfusion of the fetal periphery with a consequent accumulation of metabolic acids.

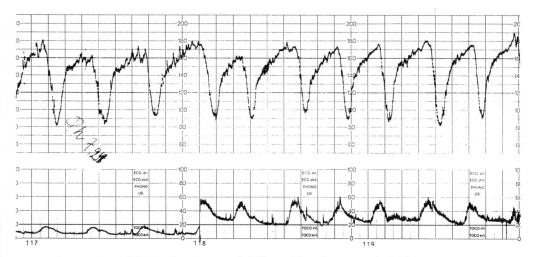

Fig. 9.9 Variable decelerations each followed by a longer period of compensatory tachycardia. Fetal scalp pH was 7.24. Spontaneous delivery occurred 20 min later and the baby was healthy.

Predictive value

The risk of developing acidosis may be suggested by the frequency, amplitude, duration, and shape of the variable decelerations, together with the variability and the baseline rate between contractions. Thus, an abrupt fall and return in FHR during each contraction, if not too frequent, and with evidence of normal placental function from a normal FHR with accelerations and variability between contractions, may be tolerated for many hours with no significant development of fetal acidaemia. However, the unpredictability of variable decelerations (in contrast to late decelerations) should always be kept in mind and fetal blood sampling should be performed according to the clinical circumstances.

Size

Variable decelerations have been graded, according to amplitude and duration, as mild, moderate, or pronounced (Kubli *et al.* 1969) (Table 9.1). Fetal scalp pH correlated with the classification such that the mean pH with mild decelerations was 7.29, with moderate decelerations was 7.26, and with pronounced decelerations was 7.15. A shortcoming of this classification is the disregard for baseline frequency from which the decelerations start. In cases of tachycardia, the amplitude and the loss of beats will be larger than with decelerations from a normal baseline. An alternative method of grading variable decelerations is by the number of lost beats (Fig. 9.10).

Table 9.1. Grading of variable decelerations

Mild	Moderate	Pronounced
Above 80 bpm, less than 30 s	Less than 70 bpm, between 30 and 60 s	Less than 70 bpm, longer than 60 s
70–80 bpm, less than 60 s	70–80 bpm, longer than 60 s	

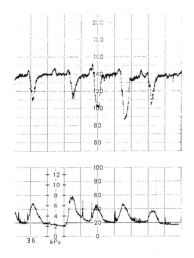

A

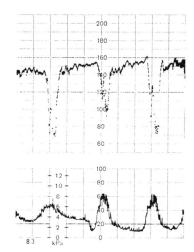

B

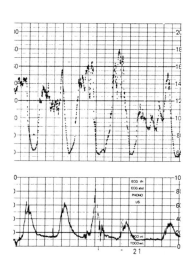

C

Fig. 9.10 Grading of variable decelerations. A, Mild—small (seldom reaching 80 bpm) and short duration; B, moderate—although the troughs rarely reach 70 bpm, the decelerations last between 45 and 60 s with a loss of between 30 and 60 bpm; C, pronounced—mostly falling below 70 bpm with a loss of more than 60 bpm and lasting more than 60 s.

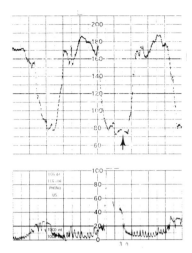

Fig. 9.11 Pronounced variable decelerations followed by periods of compensatory tachycardia. The arrow indicates a short episode of complete fetal asystole.

Many authors are of the opinion that the amplitude is of minor importance and that it is the duration and, to some extent, the shape of the variable deceleration, as well as the baseline rate and variability, that more reliably indicate fetal tolerance of variable decelerations. Duration has been shown to be important in fetal monkeys, where the longer the cord was clamped after 40 s the more severe were the defects in the fetal heart conductive system. Severe vagal stimulation may lead to virtual cardiac asystole for short periods of time (Fig. 9.11).

Shape

Variable decelerations may be grouped by shape into V-shaped (less than 60 s) (Fig. 9.12) and U-shaped (longer than 60 s) (Fig. 9.13). The V-shaped variable decelerations are generally innocent, and the fetal heart frequency returns to normal immediately after. Such decelerations may exist for several hours without development of fetal hypoxia.

In contrast to uniform decelerations, analysing the shape of variable decelerations is useful. The appearance of the decelerations gives information about the fetal condition. Atypical variable decelerations may have one or more of following criteria (Fig. 9.14).

Absence of accelerations (Fig. 9.14, 1 and 3)

Loss of initial or accompanying accelerations may often be seen in late first stage of labour, especially if variable decelerations have existed for some hours.

Variable decelerations

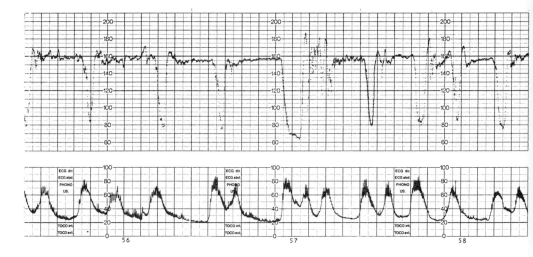

Fig. 9.12 V-shaped variable decelerations.

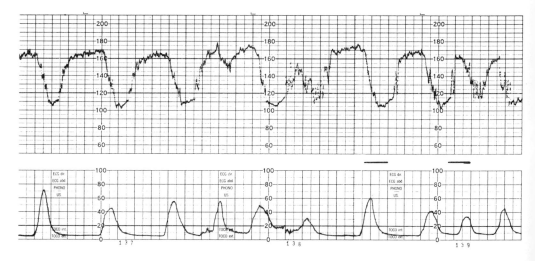

Fig. 9.13 U-shaped variable decelerations.

Animal experiments have shown that the initial acceleration is deleted in partial clamping of the cord if the fetus is hypoxic (Fig. 9.15). Initial or accompanying accelerations are a reassuring feature, even if pronounced (Fig. 9.16), and may occur either before or concomitant with the variable deceleration if they have evolved from periodic accelerations.

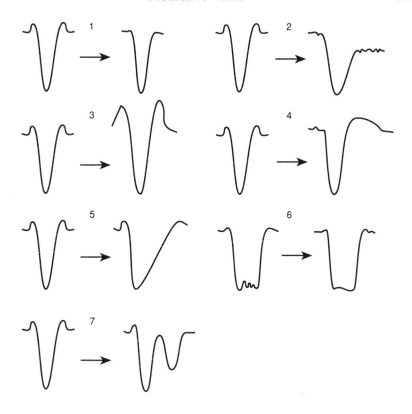

Fig. 9.14 A schematic diagram of atypical variable decelerations.

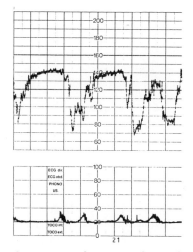

Fig. 9.15 Variable decelerations without initial or subsequent accelerations.

Variable decelerations

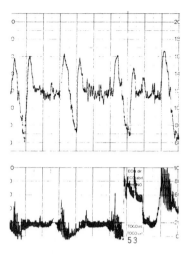

Fig. 9.16 Pronounced initial and subsequent accelerations (as though single accelerations with contractions have been interrupted by shorter-lasting variable decelerations towards the peak of the contraction).

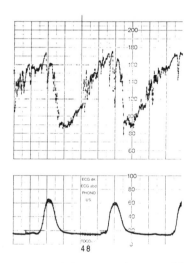

Fig. 9.17 Slow recovery of the FHR after variable decelerations.

Slow recovery (Fig. 9.14, 5)

A slow return to the baseline after the deceleration (Fig. 9.17) is usually reassuring if variability remains present. Although this is often seen with frequent contractions when the fetal heart rate has no time to 'recover', this pattern may evolve into a prolonged deceleration and be associated with fetal hypoxia.

Predictive value 165

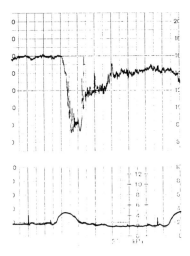

Fig. 9.18 The FHR remained lower after this variable deceleration.

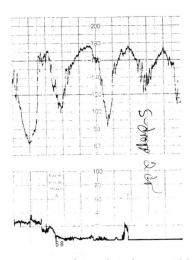

Fig. 9.19 Compensatory tachycardia after variable decelerations.

Lower baseline (Fig. 9.14, 2)

The baseline may remain at a lower level (Fig. 9.18) after the deceleration if vagal stimulation is prolonged. This could well be associated with fetal hypoxia but may also be the result of vagal stimulation by other means, such as rapid descent of the fetal head through the birth canal.

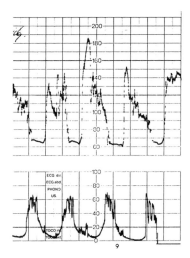

Fig. 9.20 Loss of variability during variable decelerations.

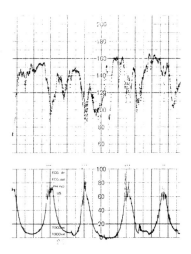

Fig. 9.21 Biphasic decelerations (comprising early and late components).

Rebound tachycardia (Fig. 9.14, 4)

A 'rebound' tachycardia (Fig. 9.19) should be regarded as a warning signal of fetal hypoxia. The baseline FHR may settle at a higher level after each contraction. Loss of short-term variability and a rounded, blunt, acceleration (overshoot) is often an ominous sign. In our experience, the frequency of asphyxia is particularly high among premature babies showing this type of pattern.

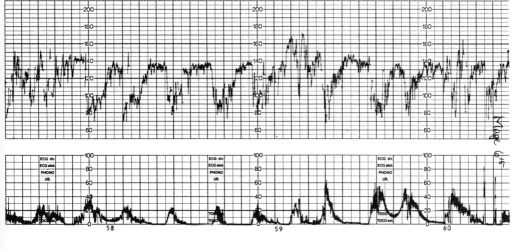

Fig. 9.22 Uterine activity monitored by external tocography. Variable decelerations and high variability. Normal delivery and neonatal condition 2 h later.

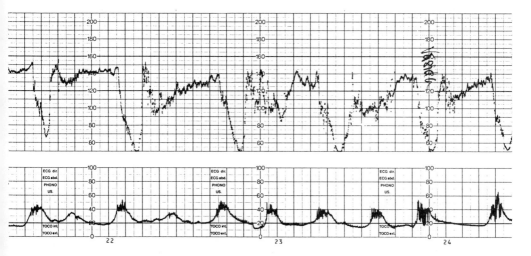

Fig. 9.23 Pronounced variable decelerations with changes in baseline FHR between contractions.

Loss of variability (Fig. 9.14, 6)

Loss of variability during the deceleration (Fig. 9.20) may be associated with progressive fetal acidosis, especially with prolonged, pronounced variable decelerations (falling to between 50 and 60 bpm).

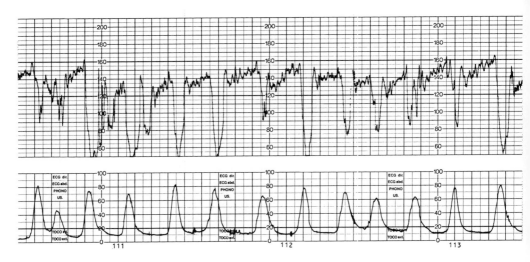

Fig. 9.24 Moderate to pronounced variable decelerations with normal FHR and variability easily seen between decelerations.

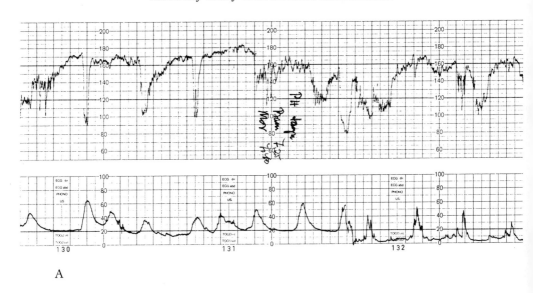

A

Fig. 9.25 (caption opposite).

Biphasic decelerations (Fig. 9.14, 7)

Biphasic variable decelerations or combined decelerations (Fig. 9.21) are sometimes referred to as variable decelerations with a late component (see Chapter 10).

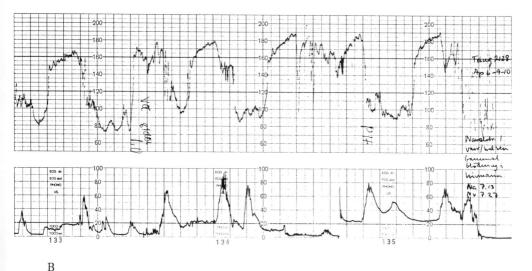

B

Fig. 9.25 A multipara at 40 weeks after oxytocin augmentation because of prolonged, prelabour, rupture of the membranes. Her previous delivery had been an emergency Caesarean section for failure to progress. A, The cervix was fully dilated after 9 h of labour and the fetal head was 1 cm below the spines. Complicated variable decelerations with slow recovery and a compensatory tachycardia. Fetal scalp pH at mark 131 was 7.25; B, the CTG continued from A. A baseline tachycardia became apparent and the decelerations became longer with absent variability. A fetal blood sample was performed (35 min after the previous one) and the pH was 7.10. Delivery was expedited by forceps and Apgar scores were 6 and 9. The cord artery pH was 7.13 and the vein pH was 7.27. The umbilical cord was found wrapped once around the baby's neck. Note the rapid fall in pH associated with atypical variable decelerations and an abnormal baseline FHR.

Baseline frequency and variability

Evaluation of amplitude, duration, and shape of the variable deceleration gives substantial information, but the most important aspect of the CTG for assessment of fetal well-being is the baseline rate and variability between the decelerations. Variable decelerations often make the FHR look confusing and difficult to interpret. Variable decelerations, associated with high variability, make the assessment of variability and accelerations even more difficult. Furthermore, the deceleration is often combined with an acceleration that is hard to distinguish from accelerations between decelerations. An associated reactive FHR pattern may be difficult to identify (Figs 9.22 and 9.23).

Variability should be assessed from the baseline between the variable decelerations. This is easy when the baseline is well defined and the intervals

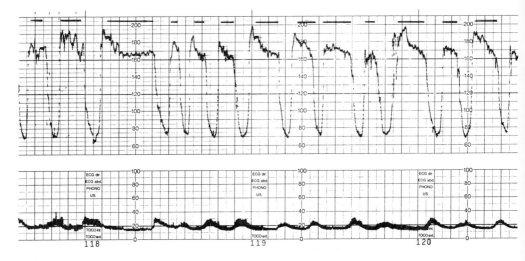

Fig. 9.26 Assessment of the baseline (periods marked at top of trace) between pronounced variable decelerations.

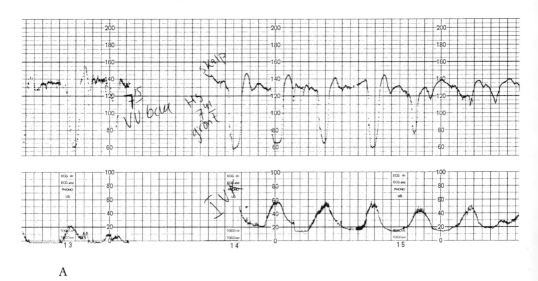

A

Fig. 9.27 (caption opposite).

between decelerations are sufficient (Fig. 9.24). Difficulties arise when the baseline alters between one deceleration and the next due to a tachycardia or a slow return to baseline (Fig. 9.25). With frequent variable decelerations the intervals between are too short to make an adequate assessment of variability, and the true baseline may be difficult to determine (Figs 9.25 and 9.26).

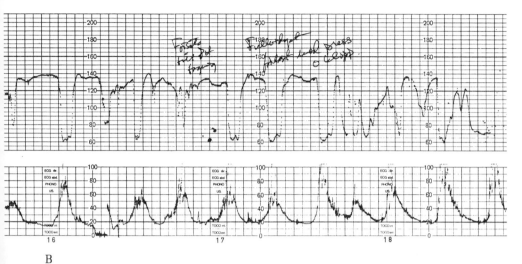

B

Fig. 9.27 A multipara at 39 weeks in spontaneous labour at 6 cm dilatation. A, A poor quality Admission Test, with excessive signal loss using external transducers, suggested large variable decelerations. The woman was moved into a delivery room and the membranes ruptured for internal monitoring. Thick meconium-stained liquor was seen. Direct FHR and intrauterine pressure records showed recurrent moderate variable decelerations, each followed by baseline overshoot. The baseline showed no variability; B, the trace continued until delivery 30 min later. Fetal blood sampling was unsuccessful at mark 17 but the woman was now fully dilated and pushing. Rapid vaginal delivery was expedited by episiotomy. Apgar scores were 0, 3, and 4 at 1, 5, and 10 min respectively.

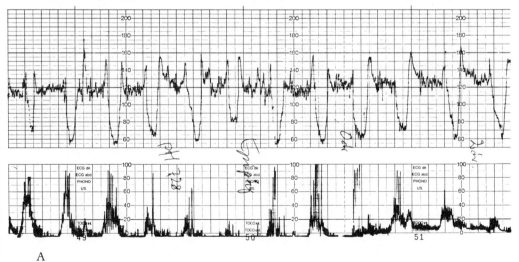

A

Fig. 9.28 (caption overleaf).

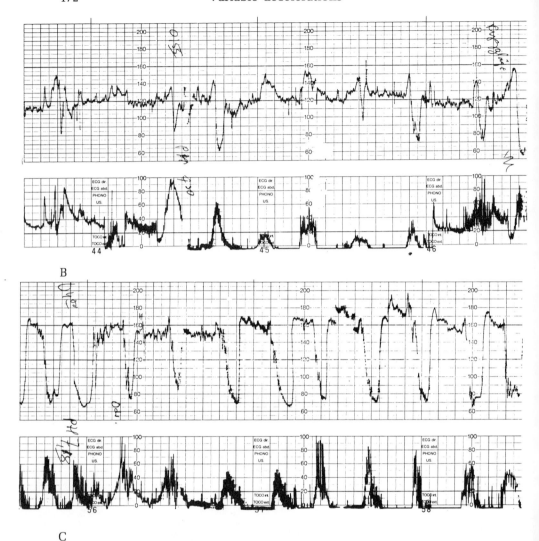

C

Fig. 9.28 A multipara at 39 weeks in spontaneous labour at 6 cm after rupture of the membranes. A, Occasional variable decelerations resulted in a fetal blood sample (at mark 44), which was normal (7.30). Labour was augmented with oxytocin (2–4 mU/min); B, the cervix was now 9 cm with the head just above the spines. Pronounced variable decelerations had developed but the normal baseline, variability and accelerations indicate a healthy fetus. C, 50 min later, the cervix was fully dilated. The baseline rate had risen and the variability was lower. A fetal blood sample (performed at mark 56) was mildly acidaemic (pH 7.18) and a vacuum extraction was performed 10 min after the end of the trace. Apgar scores were 8 and 9. The umbilical artery pH was 7.15 and the vein was 7.23. The cord was looped once around the baby's neck. Such acidaemia is usually of respiratory type if related to cord compression.

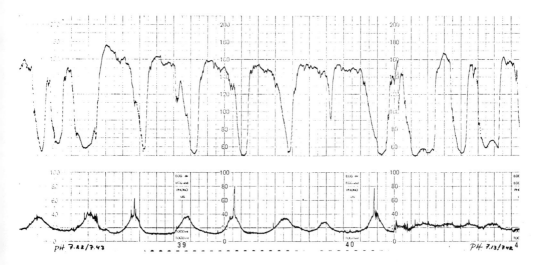

Fig. 9.29 FHR and external record of uterine activity. The baseline was slightly raised at between 150 and 165 bpm with no variability. Most variable decelerations were followed by overshoot of the baseline. Fetal scalp pH at the beginning of the record was 7.22 but was 7.13 within 30 min. After an emergency Caesarean section, Apgar scores were 5 and 7, and the umbilical artery pH was 7.11.

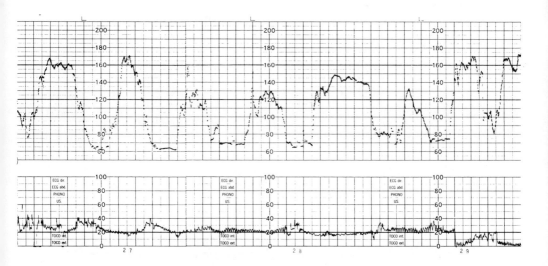

Fig. 9.30 An ominous FHR pattern with pronounced variable decelerations of prolonged duration. There was no variability during the deceleration. Immediate delivery was appropriate.

Variable decelerations

Many studies point out that loss of FHR variability (Fig. 9.27) and tachycardia between variable decelerations are important predictors of fetal distress (Fig. 9.28). Acidosis may develop rapidly with these FHR patterns. Figure 9.29 illustrates progressive fetal acidosis in a case where the fetal scalp pH dropped from 7.22 to 7.13 in less than 30 min. The CTG record shows pronounced variable decelerations with absent variability and tachycardia. Figure 9.30 shows prolonged variable decelerations with absent variability during the episodes of bradycardia.

Management

Most variable decelerations are not associated with fetal hypoxia. They do not fall below 80 bpm or have a loss of more than 60 beats but are V-shaped for less than 60 s without atypical characteristics. The baseline rate remains normal with normal variability and reactivity between decelerations (Fig. 9.31).

Relief of cord compression

In records with pronounced variable decelerations where variability, reactivity, and baseline are also affected it is necessary to assess fetal tolerance of the presumed cord compression. It is well known from clinical experience that changes in maternal position may alleviate variable decelerations and should be tried; the supine position should be avoided. The pattern may resolve if the mother is repositioned 'on all fours' (Figs 9.32 and 9.33). It may also be necessary

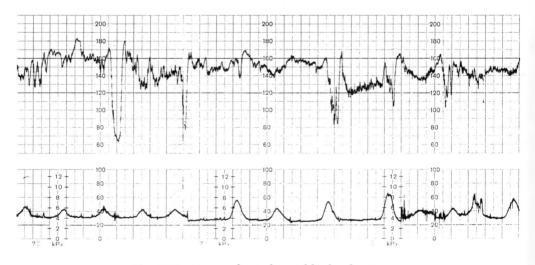

Fig. 9.31 Uncomplicated variable decelerations.

Management

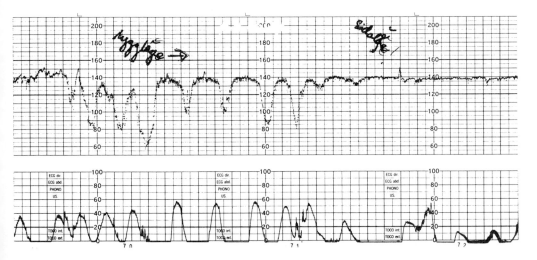

Fig. 9.32 Variable decelerations disappear after turning the mother from supine to the lateral position.

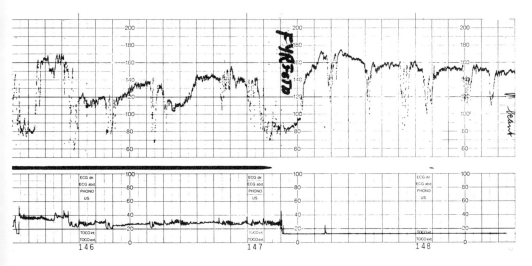

Fig. 9.33 Pronounced variable decelerations disappear after positioning the mother on her chest and knees.

to deliver the mother in positions other than the traditional supine one. Changes in position to relieve cord compression are arbitrary if the exact position of the cord is unknown (Fig. 9.34).

The sudden appearance of variable decelerations suggests the possibility of umbilical cord prolapse, which must be ruled out by vaginal examination. After

Variable decelerations

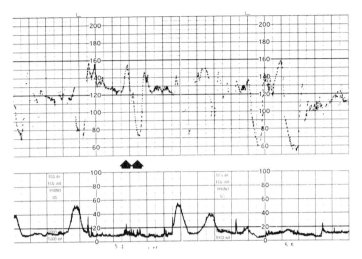

Fig. 9.34 Variable decelerations associated with fetal movements between contractions (marked with arrows). The umbilical cord was found wrapped around one foot at delivery.

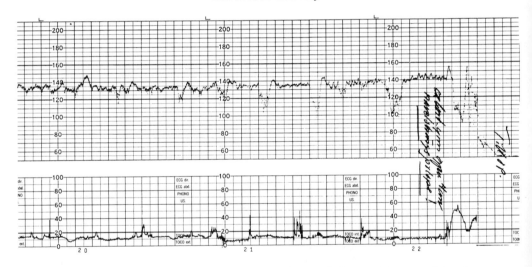

Fig. 9.35 The CTG after cord prolapse with a cephalic position, resulting in a fetal bradycardia at the end of the record.

rupture of the membranes, a prolapsed cord is often so effectively compressed that the fetal response is an acute bradycardia or prolonged deceleration rather than variable decelerations. Figure 17.6 illustrates an admission test with cord prolapse in connection with spontaneous rupture of the membranes during the record. Figure 9.35 shows a cord prolapse with vertex presentation.

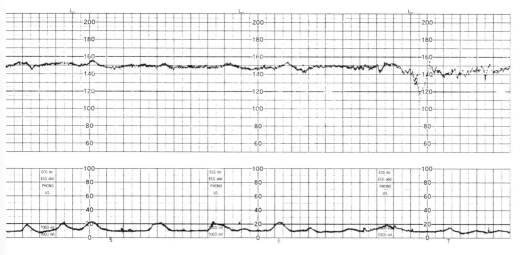

Fig. 9.36 A normal CTG despite cord prolapse with a (footling) breech position.

With breech presentations, the FHR may not indicate cord prolapse in the absence of compression, such as with a footling breech (Fig. 9.36).

Recent reports of amnioinfusion have shown this to be of value in reducing the incidence of variable decelerations after rupture of the membranes. Oligohydramnios is an important factor associated with variable decelerations both antenatally and during labour. Animal experiments have shown that variable decelerations will appear if amniotic fluid is reduced, and will disappear again when fluid is restored. Pronounced variable decelerations or prolonged decelerations may be reduced or even abolished if saline is infused through the catheter used for intrauterine pressure recording. Between 10 and 15 ml/min, up to 600–800 ml, can usually be installed until decelerations have disappeared.

Traditionally, oxygen has been administered to the mother. Administration of atropine has been tried in the past and was successful in abolishing variable decelerations. However, this does not ameliorate the fetal situation but disguises the diagnosis and interferes with the fetal response, which is likely to be a protective, rather than harmful, adaptation.

Fetal blood sampling

Assessment of fetal pH is necessary if hypoxia is suspected. Blood sampling should ideally be done with the mother in the lateral position and between decelerations. Oxytocin infusion should be stopped at least 15 min before sampling. However, an FHR record with tachycardia and loss of variability indicating progressive fetal acidosis is an indication for an expedient delivery rather than scalp blood sampling (Figs 9.27, 9.29 and 9.30).

Slow progress

Decelerations often appear in late first stage of labour after several hours of contractions. An intravenous infusion of between 500 and 1000 ml fluid to the mother in this stage of labour might improve the fetal situation.

Administration of oxytocin infusion for augmentation of activity usually aggravates the CTG record. A difficult situation arises when the spontaneous uterine activity is insufficient for progress to cervical dilatation and yet administration of oxytocin results in a significantly worse FHR appearance.

Fetal position and presentation

Occiput posterior position

Fetal position and presentation are also related to variable decelerations. Occiput posterior (OP) position occurs in 10–15 per cent of all labours and spontaneous rotation to occiput anterior (OA) position will take place in two-thirds of all cases. The OP position is associated with dysfunctional labour and an abnormal contraction pattern.

We investigated the FHR patterns in 123 mothers where the babies were born OP and matched them with 123 OA deliveries. No difference was seen in baseline frequency or variability. Variable decelerations were, however, significantly more frequent in OP labours, appearing in 59.4 per cent, compared with 32.8 per cent in the OA group. Early and late decelerations appeared equally in both

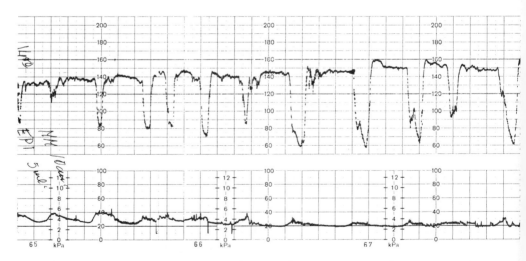

Fig. 9.37 Variable decelerations without accelerations in a case of occiput posterior position.

Table 9.2. Fetal head position and depth of variable decelerations

Loss of beats	Position	
	OP (%) (n = 73)	OA (%) (n = 43)
Less than 30 bpm,	16	44
30–60 bpm,	49	44
more than 60 bpm	35	12

groups. Absence of the initial acceleration was twice as common in the OP group than in the OA group (Fig. 9.37). Cord complications seen at delivery were found in about 30 per cent of both groups. The variable decelerations were also more pronounced in the OP than in the OA group (Table 9.2). The babies were born in good condition in both groups. Despite pronounced variable decelerations, the frequency of fetal acidaemia was not higher in the OP group.

In fetal sheep experiments, 'variable' decelerations have been produced by compression of the fetal head or by compression of the eyes. The abnormal presentation of fetal head when deflexed seems, therefore, to produce a similar effect. This suggestion is also supported by the finding of a high incidence of variable decelerations in labours with a face presentation.

Breech presentation

A fetus in breech presentation calls for special monitoring. Obstetricians reluctant to rupture the membranes will only have the option of external monitoring without fetal blood pH analysis. Although not impossible, it is technically more difficult to perform fetal blood sampling from the foot or breech. Several studies have shown that a reduction in fetal pH (of metabolic type) may occur during the second stage, particularly if protracted, increasing the risk for asphyxia at birth.

In breech presentation there is a higher frequency of variable decelerations and high variability compared with vertex presentations. The reason may be the increased risk for cord compression. In addition, there is a predisposition for cord compression and the umbilical cord is often shorter in breeches. We compared the FHR patterns from 56 breeches with records from fetuses in the OA or OP position. Pronounced variable decelerations, with a loss of more than 60 beats, were more frequent with breech presentation (76 per cent) than with OP. However, low Apgar scores (less than 7) at 1 min were only associated with variable decelerations if variability was decreased or absent (Fig. 9.38).

Variable decelerations

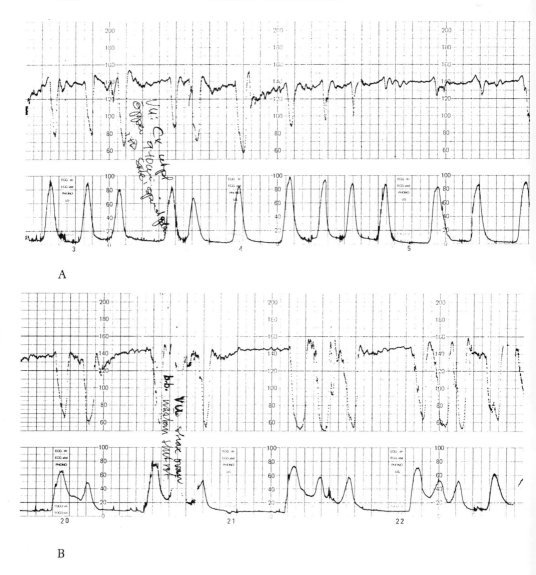

Fig. 9.38 (caption opposite).

Summary

Variable decelerations are the most common variety of FHR deceleration found in labour. Compression of the umbilical cord, resulting in a transient episode of fetal hypoxia, is believed to be primarily responsible for this response.

Summary

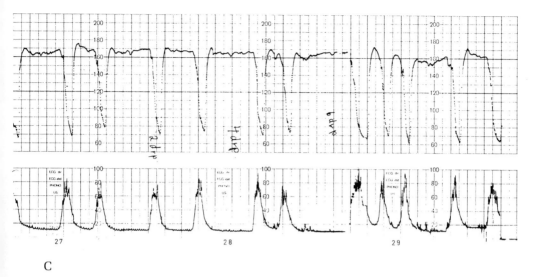

C

Fig. 9.38 A primipara in spontaneous labour with a breech presentation at 38 weeks. A, The breech was at the spines at 9 cm dilated. Mild and moderate variable decelerations and low baseline variability; B, 2 h later, the breech is 2 cm below the spines. The variable decelerations are pronounced. Fetal pH was normal (7.31); C, 50 min later, the last 30 min of labour. A fetal tachycardia with absent variability had developed. The variable decelerations remained pronounced. The baby was in good condition with Apgar scores of 7 and 8 (compare with Figure 6.10).

Evidence of sustained fetal hypoxia is inferred from atypical features related to a slowing of the reflex FHR response. Alterations in the baseline FHR between decelerations are also important with respect to assessment of general fetal condition.

10
Combined decelerations

Introduction

A combined deceleration has two components, which appear as a W waveform, and represent the combination of an early or variable deceleration and a late

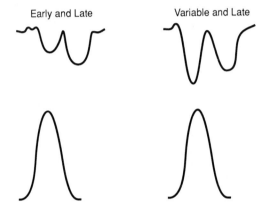

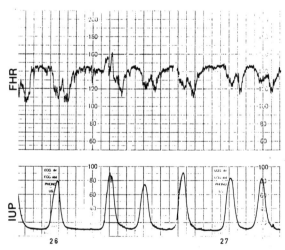

Fig. 10.1 A schematic figure of the two combinations of combined decelerations.

Fig. 10.2 Combination of early and late decelerations associated with high intrauterine pressure during contractions.

Introduction

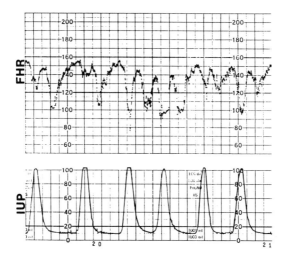

Fig. 10.3 Combination of variable and late decelerations associated with very high intrauterine contraction pressures.

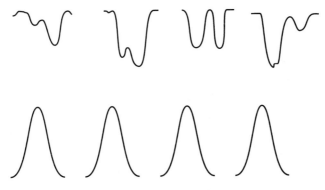

Fig. 10.4 A schematic figure showing variations of the W waveform pattern of combined decelerations.

deceleration in response to a single contraction (Figs 10.1, 10.2, 10.3 and 10.4). Such decelerations have been described as biphasic, or variable with a late component.

Aetiology

Combined decelerations, present for at least 30 min during the first stage of labour, were found in 70 labours (1.1 per cent) during a 2-year period in Lund. They were found predominantly late in the first stage of labour, mainly associated with infusion of oxytocin. Of 40 cases that occurred after the cervix had dilated more than 5 cm, more than half were after the cervix was dilated 9 cm or more. Only 5 of the 70 cases were spontaneous labours (Table 10.1). Uterine activity, recorded by intrauterine pressure catheter in 47 cases, was abnormal in 37 (78.7 per cent) (Table 10.2), and only one of these was a spontaneous labour.

Table 10.1. Uterine activity associated with combined decelerations

	Total	Percentage
Spontaneous	5	7.1
Oxytocin	65	92.9
induced	16	
augmented	49	

Table 10.2. Quantitative analysis of uterine activity in 47 cases of combined decelerations

	Number	Percentage
Normal	10	21.3
Abnormal	37	78.7
Montevideo units $\geqslant 250$	32	
basal tone $\geqslant 20$ mmHg	11	
contraction amplitude $\geqslant 70$ mmHg	18	
frequency $\geqslant 5$ per 10 min	13	
tetanic	11	

Another indication of the predominant association between variable decelerations and oxytocin 'overstimulation' was that the late component of the deceleration usually appeared simultaneously with the onset of abnormal uterine activity (Fig. 10.5). In addition, whereas the late component disappeared when the oxytocin infusion was stopped, the first component (early or variable deceleration) persisted.

Overstimulation by oxytocin is the most common cause of uterine hyperactivity. Different types of FHR pattern may occur with abnormal labour, as discussed in Chapter 4. Overstimulation may produce transient fetal bradycardia, prolonged or late decelerations, or combined decelerations as a result of abnormal uterine activity. The late component of these responses reflects the possibility

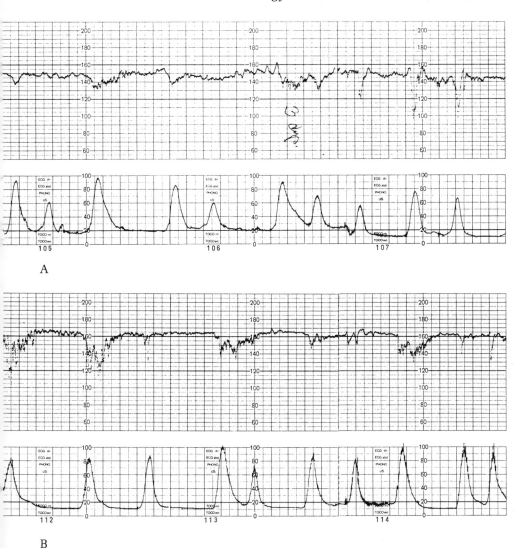

Fig. 10.5 A primipara at 40 weeks gestation with slow progress in spontaneous labour. The CTG is by direct FHR and intrauterine pressure records. A, Cervical dilatation was 8 cm with the fetal head at the level of the ischial spines. A low dose of oxytocin (between 2 and 3 mU/min) had just been started. Baseline FHR was 150 bpm with low variability. Small early and variable decelerations were present; B, 50 min later, at the same rate of oxytocin infusion, the contraction amplitudes were higher and mild combined decelerations had developed. Scalp blood pH was 7.36. In the second stage of labour there were more pronounced combined decelerations (variable and late). The Apgar scores were 6 and 9. The umbilical artery pH was 7.19. The cord was looped twice around the fetal neck.

of direct myocardial hypoxic depression due to excessive interruption of fetal oxygenation (Fig. 10.6). Management of cases with combined decelerations is, therefore, similar to other circumstances when abnormal uterine activity results in late decelerations (Figs 10.7, 10.8, 10.9 and 10.10).

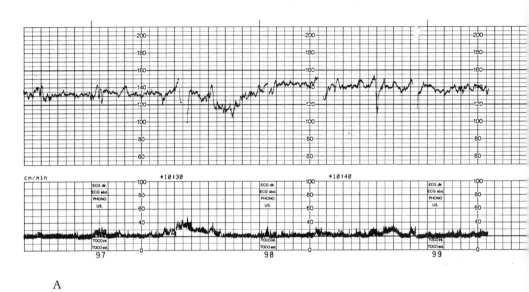

A

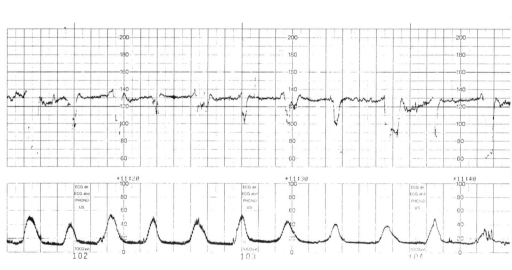

B

Fig. 10.6 (caption opposite).

Aetiology

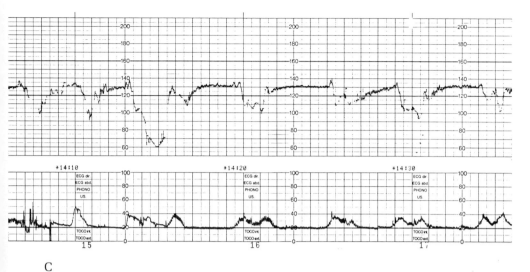

C

Fig. 10.6 A primipara at term with hypertension in pregnancy. A, Prior to induction of labour by intracervical prostaglandin gel, the FHR was normal; B, 30 min after administration of prostaglandin gel, regular contractions are evident. The baseline FHR was 130 bpm with low variability. Small, variable decelerations were seen and, occasionally, a late component was present; C, 3.5 h later, regular contractions with combined decelerations, were present. Delivery occurred 7 h later and the Apgar scores were normal. Birth weight was 2480 g. A 2 cm by 6 cm retroplacental clot was found.

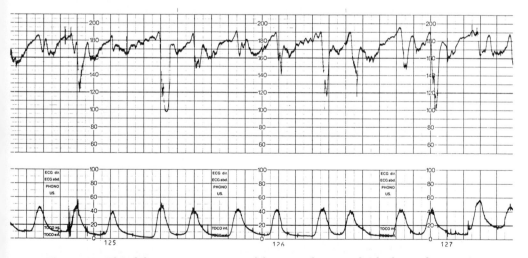

Fig. 10.7 This labour was augmented by a moderate-to-high dose of oxytocin (30 mU/min). The FHR appears ominous, with a tachycardia, low variability, and combined decelerations (variable and late). The external record of uterine activity provided no evidence of overstimulation.

Combined decelerations

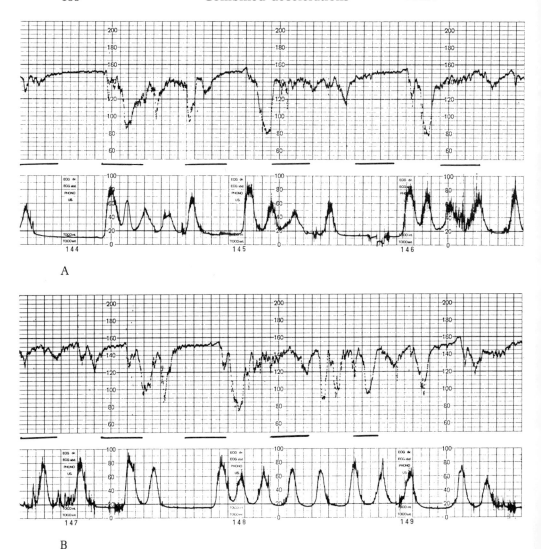

Fig. 10.8 A primipara at term with spontaneous rupture of the membranes. The cervix was dilated only 1 cm. An oxytocin infusion was commenced in order to induce labour. A, After 8 h of oxytocin infusion, abnormal uterine activity, with multiple coupling of contractions, was seen on the intrauterine pressure record. Baseline FHR was 150 bpm with low variability and combined variable and late decelerations. Fetal scalp blood pH was 7.26; B, the same CTG continued. The combined decelerations were more pronounced. Repeated scalp blood pH was 7.21. The cervix was effaced and dilated between 2 and 3 cm. An emergency Caesarean section was performed and the Apgar scores were normal. The oxytocin infusion had not been stopped before scalp blood sampling.

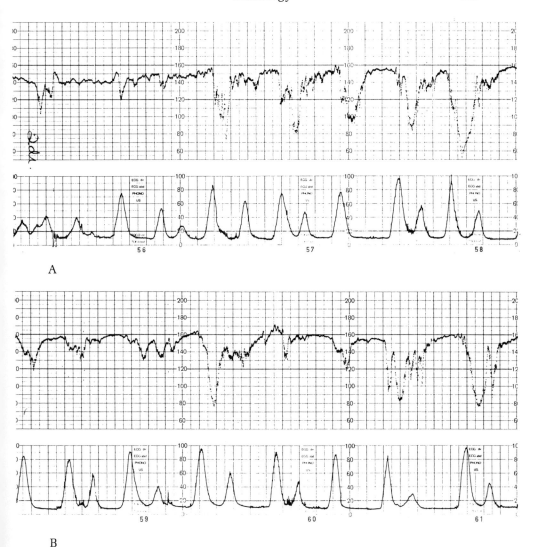

Fig. 10.9 A primipara at term who required induction of labour because of pre-eclampsia and a growth-retarded fetus. Meconium-stained liquor was seen when the membranes were ruptured. A, After 1.5 h, intrauterine pressure during contractions was high. Baseline FHR was between 140 and 155 bpm with low variability. Combined variable and late decelerations were present despite an oxytocin infusion rate of only 2 mU/min; B, the same CTG continued. Baseline FHR was between 155 and 160 bpm with a reduction in variability. Recurrent combined decelerations persisted. An emergency Caesarean section was performed for suspected fetal distress. Apgar scores were 4, 7, and 8 at 1, 5, and 10 min. The umbilical artery pH was 7.17 and the vein pH was 7.28. Birthweight was 2330 g.

Combined decelerations

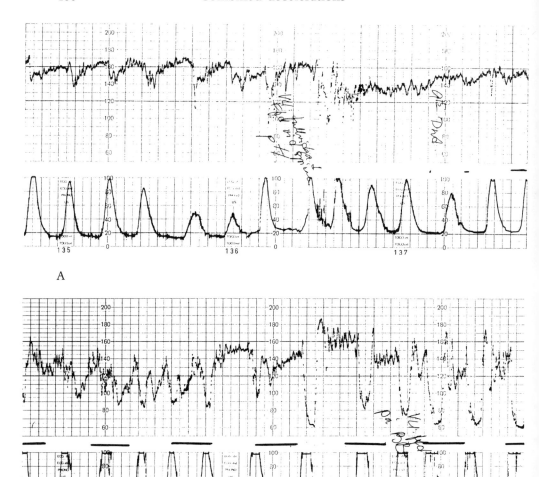

Fig. 10.10 A multipara at 41 weeks with spontaneous rupture of the membranes and weak contractions. Augmentation with oxytocin infusion was commenced. A, Advanced first stage of labour (cervix dilated 9 cm). The intrauterine pressure record showed overstimulation of the uterus. Small combined decelerations can be seen. The scalp blood pH was normal (7.42). The oxytocin infusion continued; B, 30 min later, pronounced combined decelerations developed. Delivery occurred after a further 30 min without a change in the FHR pattern. The Apgar scores were 6, 8, and 10 at 1, 5, and 10 min. The cord was looped twice around the neck of the fetus. Reduction of the oxytocin infusion rate during the last two hours would have been advisable.

Augmentation of labour with oxytocin infusion is an integral part of the management of slow labour. Difficulties arise with individual variations in response to oxytocin, and the uterus becomes increasingly more sensitive to oxytocin as labour progresses. Descent of the fetal head also stimulates the uterus through the Ferguson reflex. Overstimulation during this part of labour is, therefore, quite likely if the dose of oxytocin required to initiate progress is maintained throughout the first stage once progress is established.

Summary

The FHR should be continuously monitored if there is a risk of uterine hyperactivity. Combined decelerations serve to warn of the possibility of excessive oxytocin administration if uterine activity is monitored by external means. The oxytocin infusion should be reduced or stopped when combined decelerations appear. Intrauterine pressure recording should then be considered. Scalp blood sampling is indicated if the combined decelerations persist or become pronounced (Fig. 10.11).

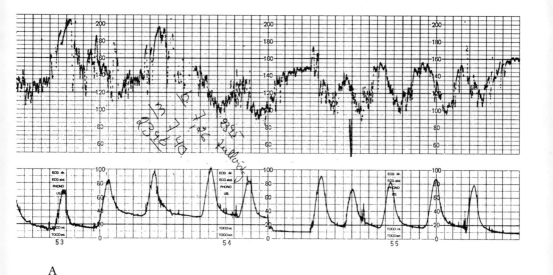

Fig. 10.11 (caption overleaf).

Combined decelerations

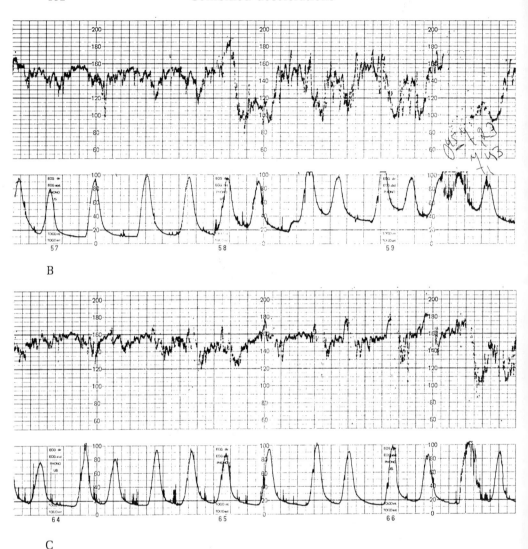

Fig. 10.11 A multipara whose previous delivery was by Caesarean section. Induction of labour was advocated because of prolonged pregnancy (42 weeks). After 13 h the cervix was dilated 8 cm. A, Oxytocin was being given at 5 mU/min. Uterine activity was monitored by an intrauterine pressure catheter. Baseline FHR was 150 bpm with normal variability. Mild to moderate combined decelerations were present. Scalp blood pH was 7.26; B, 10 min later, abnormal uterine activity developed with increased basal tone. The FHR was 155 bpm with normal variability and combined decelerations as before. Fetal scalp blood pH was now 7.23. The combined decelerations disappeared and the basal tone became normal when the oxytocin infusion was stopped.

11
Prolonged decelerations

Introduction

There is no established definition for a prolonged deceleration. We consider an FHR below 100 bpm for more than 3 min, or below 80 for more than 2 min, to be prolonged. Most prolonged decelerations last longer than this and may also be considered as transient episodes of bradycardia. Prolonged decelerations are usually sporadic and are often related to obstetric events or interventions. Corrective measures will usually solve the problem. Prolonged decelerations may be considered innocuous if recovery occurs within 10 min. Occasionally, such decelerations are an indication of an obstetric catastrophe, such as rupture of the uterus or uterine scar, abruption of the placenta, rupture of vasa praevia and fetal bleeding, or cord prolapse. However, in such cases recovery of the FHR is unlikely and the bradycardia becomes prolonged.

Aetiology

Abnormal uterine activity

Excessive uterine activity reduces the maternal placental circulation and, if severe, may compromise maternal–fetal exchange in an otherwise healthy placenta. If placental reserve is already reduced, as is likely with pre-eclampsia and as is evident with intrauterine growth retardation, a reduction in fetal oxygenation may occur, even with normal uterine activity. Although abnormal uterine activity may be seen in spontaneous labour with normal progress, it is most commonly a result of oxytocin or prostaglandin. Prolonged decelerations in such cases are often preceded by baseline changes or other decelerations.

Uterine hyperstimulation resulting from oxytocin will usually resolve spontaneously within a few minutes of stopping the infusion. After prostaglandin treatment, or with spontaneous hyperactivity, a β-receptor agonist (e.g. salbutamol or terbutaline) can be given as an infusion or bolus injection. This will relax the uterus and improve fetal oxygen supply, resulting in recovery of the FHR (see Figs 4.20 and 11.1).

Abnormal uterine activity may also occur spontaneously during the antenatal period (see Figs 11.2, 14.22 and 14.23) and during labour (see Fig. 4.8). An

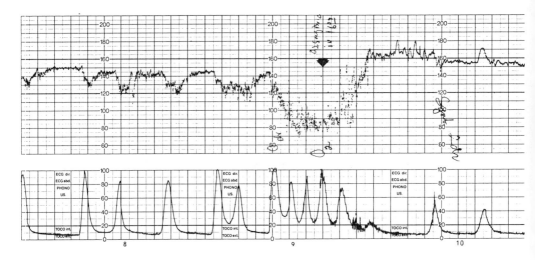

Fig. 11.1 Uterine activity monitored by intrauterine pressure. The FHR is between 145 and 150 bpm with normal variability and combined decelerations. At mark 9, a prolonged deceleration resulted from the abnormal uterine activity with increased basal tone. The infusion of oxytocin was stopped and terbutaline (0.25 mg i.v.) was given (at the arrow). The uterus relaxed within 2 min and the FHR recovered.

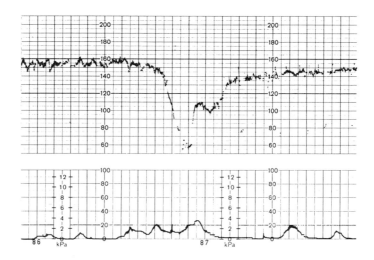

Fig. 11.2 A CTG in late pregnancy with spontaneous hyperactivity of the uterus producing a prolonged deceleration.

Aetiology

abnormal contraction, lasting for several minutes, will provoke a prolonged deceleration, which can be regarded as a physiological response by the fetus. The transient episode of fetal hypoxia will not seriously affect the fetus if superimposed upon normal oxygenation and, in such circumstances, is innocuous. This is usually ascertained by a normal FHR pattern before and after the abnormal contraction (see Chapter 4).

Cord compression

Cord compression disturbs the fetal circulation and, if severe, will result in fetal bradycardia or a prolonged deceleration, depending on the duration (Fig. 11.3). If the compression is less, FHR changes may vary and can be preceded by periodic accelerations or variable decelerations. Prolonged decelerations may develop from variable decelerations, particularly with excessive uterine activity. Warning signals, such as slow recovery from the variable deceleration, may appear earlier. Mild forms of cord compression may also occur with rapid descent of the fetal head in the birth canal.

Maternal hypoxia or hypotension

Maternal seizures (eclamptic or epileptic fits), with systemic hypoxia, can provoke prolonged decelerations (Fig. 11.4). Maternal hypotension, secondary to the supine position and aortocaval compression, is the most common cause of

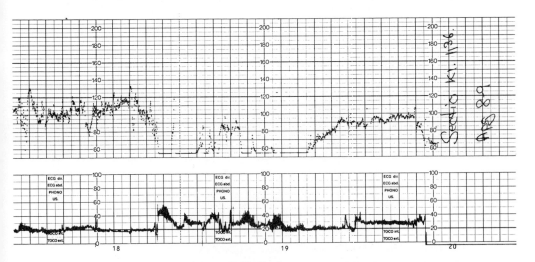

Fig. 11.3 A pronounced prolonged deceleration of the FHR below 60 bpm for which an emergency Caesarean section was performed late in the first stage of labour. Apgar scores were normal. The cord was tightly round the fetal neck three times and probably became compressed as the fetal head descended.

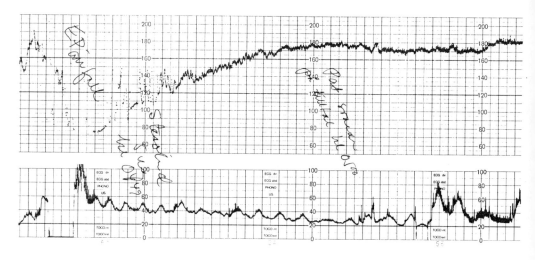

Fig. 11.4 A prolonged deceleration in association with an epileptic fit. Control was achieved by diazepam 5 mg i.v. A compensatory fetal tachycardia and reduced variability probably represent residual catecholamine release secondary to the hypoxic episode.

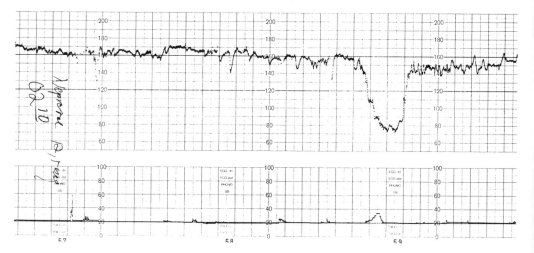

Fig. 11.5 A prolonged deceleration 15 min after injection of an antihypertensive drug in high dose (nepresol 12.5 mg i.v.), which caused transient maternal hypotension.

prolonged decelerations. These decelerations occur during the antenatal period as well as in labour. Epidural anaesthesia is a common cause of maternal hypotension (see Chapter 16) and resultant prolonged decelerations are usually preceded by a normal FHR pattern. The duration of the deceleration is often only a few minutes and recovers when the woman is moved into the lateral position.

Aetiology

Antihypertensive drugs, such as hydralazine, when given in high doses intravenously, can reduce uteroplacental circulation. A prolonged deceleration may appear, especially if the maternal systolic blood pressure falls below 80 mmHg (Fig. 11.5).

Acute fetal bleeding

Bleeding after cordocentesis or accidental puncture of a cord vessel at amniocentesis may cause prolonged decelerations (or FHR bradycardia). The rare complication of rupture of umbilical vessels, passing in front of the presenting part within the membranes (vasa praevia), may occur with artificial or spontaneous rupture of membranes during labour. The fetal response to acute haemorrhage and interruption of the umbilical circulation may be a severe prolonged deceleration or FHR bradycardia (Fig. 11.6). Vaginal bleeding is usually seen. Prompt delivery of the fetus is necessary in this circumstance.

Bleeding from the fetal scalp after blood sampling for pH is an uncommon complication but may cause FHR changes if severe. The place of incision should be inspected after sampling. It is routine in many departments to hold a cotton swab on the incision site for a short while after sampling.

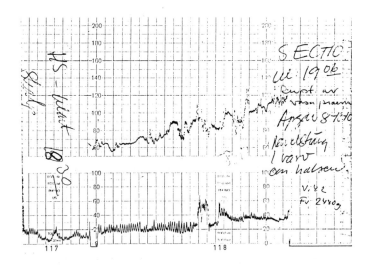

Fig. 11.6 A prolonged deceleration after artificial rupture of the membranes. Rupture of a vasa praevia was suspected when vaginal bleeding was observed. The diagnosis was verified at Caesarean section. The fetal outcome was good because action was prompt.

198 Prolonged decelerations

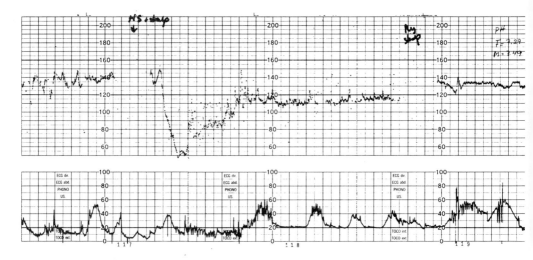

Fig. 11.7 A prolonged deceleration of artificial rupture of membranes and application of scalp electrode. The scalp blood pH was normal after recovery (at the end of the record).

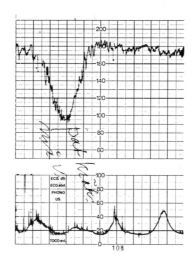

Fig. 11.8 A prolonged deceleration during maternal vomiting.

Fetal vagal activity

A prolonged deceleration may be a benign fetal vagal reflex in association with obstetric interventions like insertion of intrauterine catheter or application of a scalp electrode (Fig. 11.7). At vaginal examination, prolonged decelerations may be produced by pressure on the fetal head or, indirectly, by abnormal uterine action. Prolonged decelerations are also associated with maternal activity, such as micturition on a bedpan (see Fig. 4.8) or vomiting (Fig. 11.8), and are likely to be the result of transient interruptions in uteroplacental perfusion.

Terminal response

Prolonged decelerations can be a sign of imminent intrauterine death if preceded by an ominous FHR pattern with loss of FHR variability or a sinusoidal pattern (see Chapter 15).

Management

Most cases of prolonged deceleration of the FHR will recover within a few minutes. Routine measures include repositioning the woman in a lateral position, vaginal examination to exclude cord prolapse, cessation of oxytocin infusion, and intravenous infusion of fluid (especially after epidural block). These patterns give rise to much anxiety, particularly if they continue despite the usual corrective actions, and result in a rather rushed preparation for operative delivery. Once such preparations are in progress, usually by 10 min, it is very difficult not to proceed with operative delivery, even if the FHR then shows signs of recovering. The risk of fetal acidosis begins to rise after 5 min if the FHR has not recovered, particularly if the bradycardia is pronounced and there is loss of variability. Delivery should be expedited to avoid continuation of such a prolonged deceleration much beyond 20 min.

Fetal blood sampling

Table 11.1 shows scalp blood pH results in relation to variability and duration of bradycardia associated with prolonged decelerations in labour. Terbutaline was administered to all patients as a bolus injection between 2 and 14 min after the beginning of a deceleration. Of 33 cases with prolonged bradycardia, the FHR improved in 30 cases (two of which were cord prolapses) and 23 were

Table 11.1 FHR variability during prolonged bradycardia and scalp blood pH within 40 min of recovery after an injection of terbutaline

Duration of absent or normal variability (min)	n	Mean (range) scalp pH	Mean duration of deceleration (min)
Normal variability	13	7.32 (7.21–7.39)	5.80
Flat baseline <4 min	9	7.31 (7.22–7.40)	6.00
Flat baseline ⩾4 min	4	7.21 (7.16–7.27)	10.25

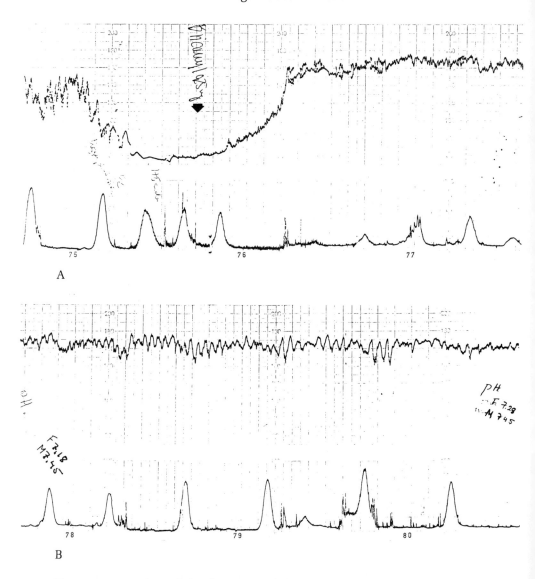

Fig. 11.9 A prolonged deceleration with fetal acidosis in a primipara in spontaneous labour after an uneventful pregnancy. The intrauterine pressure record shows normal uterine activity. There was loss of variability (flat baseline) for 6 min during the deceleration. The FHR recovered 4 min after a bolus injection (0.25 mg i.v.) of terbutaline. Scalp blood pH obtained 15 min after the end of the deceleration was 7.18. A repeat scalp blood pH 25 min later was 7.28, demonstrating the capacity for fetal recovery from intrauterine acidosis. Although uterine activity appeared normal, administration of the β-mimetic was justified by the effect on contraction rate and associated FHR recovery.

delivered vaginally. The FHR did not recover in three cases (two with manifest fetal distress and one after placental abruption following artificial rupture of membranes). Loss of variability during a prolonged deceleration, seemed to be an ominous sign. After a prolonged deceleration lasting between 7 and 14 min, with loss of variability for 4 min or more, scalp blood pH was, on average, 7.21 (measured after recovery of the FHR) (Fig. 11.9). Low variability, lasting less than 4 min, or a normal variability during the deceleration seems, to be associated with a low risk of fetal acidosis if the duration of the deceleration was 10 min or shorter (Table 11.1).

Tocolysis

Relaxation of the uterus can benefit fetal oxygenation. The fetal cardiac output is mainly dependent on a maintained heart rate and is sharply reduced during fetal bradycardia. Animal studies indicate that endogenous increase of β-receptor activity maintains normal FHR and opposes the fetal bradycardia which results from increased vagal activity, especially induced by hypoxia. β-adrenergic receptor activity maintains the FHR, and consequently cardiac output, through its inotropic action. Vasodilatation of the placental vascular bed can maintain umbilical blood flow at acceptable levels during hypoxia and probably reduces the vascular resistance of the umbilical unit. A β-receptor agonist, given in cases with bradycardia, may enhance the potential beneficial effects described.

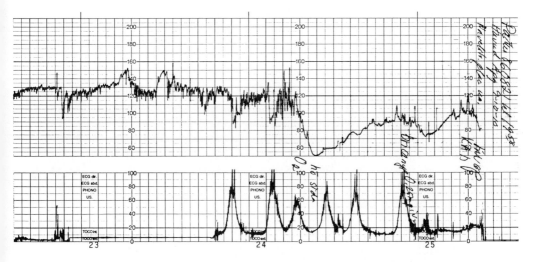

Fig. 11.10 A prolonged deceleration with loss of variability during the bradycardia. Terbutaline (0.25 mg i.v.) was given after 5 min but did not effect recovery and so an emergency Caesarean section was performed. Apgar scores were normal and no explanation for the bradycardia was found.

Table 11.2 Scalp blood pH within 40 min of a prolonged deceleration in patients after recovery after terbulatine

Duration (min)	n	Mean scalp pH (range)
2–3	–	–
4–6	19	7.32 (7.22–7.40)
7–9	5	7.29 (7.26–7.39)
10–14	4	7.22 (7.16–7.32)

Studies suggest that inhibition of uterine activity using β-receptor agonists may be indicated in selected cases with prolonged deceleration. However, these drugs should not be given if there is vaginal bleeding of unknown origin. When used, such pharmacological treatment should be regarded as a temporary measure until it is apparent that the FHR has recovered to previous baseline. Preparation for emergency delivery should be made while recovery is awaited. If the FHR has not recovered to at least 100 bpm by 10 min then intervention should be considered (Fig. 11.10), because the risk for fetal acidosis increases with the duration of the prolonged deceleration (Table 11.2). The mean scalp blood pH was 7.22 when the FHR was below 100 bpm for 10 min or more in the material described above.

The risk for uterine atony and bleeding after β-mimetic therapy is minimal after vaginal delivery but can occur at Caesarean delivery. Propranolol (a non-selective β-blocker) may be used after the cord is clamped and will immediately counteract the relaxing effect of the β-receptor agonist.

Summary

Prolonged decelerations, lasting more than 2 or 3 min, often indicate maternal hypotension and reduced uteroplacental perfusion. However, other causes of fetal hypoxia (such as cord compression) may, if prolonged, also produce a prolonged deceleration. Management involves correction of maternal hypotension, particularly if in the supine position, by turning on to the left side. If the deceleration persists beyond 5 min then preparations for delivery should begin so that, if the bradycardia continues, delivery can be achieved within 10 to 15 min of the onset.

12
Second stage

Introduction

The second stage of labour begins when the presenting part of the fetus passes through the cervix, which is therefore described as 'fully dilated'. The level of the fetal head in the birth channel (usually related to the ischial spines) at the time of diagnosis may vary considerably, depending upon parity, efficiency of labour, and epidural anaesthesia. A primipara with the fetal head at spinal level may experience between 1 and 2 h labour in the second stage before delivery. The second stage may be even longer if she has an epidural and/or there is a fetal malposition. The risk of developing fetal acidaemia in such a labour is much greater than if the second stage is short, as is likely when there is no epidural and the fetal head is on the pelvic floor at the time of diagnosis. A multipara is likely to have a much shorter second stage than a primipara.

Clinical aspects

Duration

Opinions differ with regard to the optimal duration of the second stage of labour. Some consider the length of second stage to have minimal influence on fetal condition at birth, particularly if an epidural has been used for analgesia. Others believe that the duration of the second stage should not exceed between 30 and 60 min. However, the importance of the duration of voluntary maternal effort is generally agreed to have a significant effect on the fetal condition, particularly if the mother holds her breath and pushes during the whole of each contraction (active pushing). Such efforts are usually identifiable on the toco record as flat-topped spikes during the contraction curve. Abnormal FHR patterns frequently appear after one hour of active pushing.

Neonatal condition

Apgar score

Obstetric management of the second stage is based on experience of the neonatal condition at birth. Clinical depression at birth was common after the use of large doses of narcotic analgesics during labour or after general anaesthesia for delivery. In 1953, Virginia Apgar published a scoring system to assess the need for neonatal resuscitation at 1 and 5 min after birth, and this has since been universally adopted as a means of assessing neonatal well-being. However, one disadvantage of the Apgar scoring system is that the parameters are considered to have equal weighting, which means that skin colour is scored in the same manner as heart frequency and respiratory effort.

In practice, the newborn is given an Apgar score immediately after birth. All obstetricians are well aware of the importance of muscle tone at birth—probably the most important parameter at that moment. An atonic newborn is promptly given appropriate attention before assessment of the Apgar score. Perhaps an assessment of the muscle tone and heart frequency within 30 s would be an appropriate alternative to a 1-min Apgar score as an indicator of fetal distress.

Although a 5-min Apgar score is often used as a measure of fetal distress at birth, it could be argued that it is also a measure of the efficacy of resuscitation. In addition, comparisons using the 5-min Apgar score should be judged according to the different obstetric practices of administration of analgesics and anaesthesia in labour and at delivery.

Acid–base assessment at birth

An important complement to the Apgar score when assessing neonatal condition after birth is the acid–base balance in umbilical cord blood. During the second stage, fetal blood pH falls as the result of an increase in $P\text{CO}_2$, leading to a mild respiratory acidosis. Fetal hypercapnia may be one of the stimuli for the onset of breathing at birth. The fetus may also develop a transient metabolic acidosis during crowning. At this stage of labour, fetuses with a breech presentation generally have lower pH values than those delivering head first. The blood should be sampled before the first breath. Cord arterial pH is usually at least 0.05 units below the vein value but the arterial–venous difference is greater if cord compression has occurred during labour. Umbilical arterial pH reflects fetal condition whereas umbilical venous blood indicates the ability of the placenta to correct acidosis. Blood may be sampled from a double-clamped section of cord or by puncture of the cord artery with a fine needle attached to a 2 ml syringe coated with heparin. Pulsations in the artery may continue after the puncture. In some departments umbilical cord blood acid–base balance is assessed routinely for all births.

Many factors influence the cord blood pH and, surprisingly, both low and high values are not uncommon. The mean cord artery pH value for normal deliveries is just below 7.3 and two standard deviations below this is in the range 7.1 to 7.15. The correlation with Apgar scores is not good, which indicates that the two methods are different indicators of fetal/neonatal condition. In fact,

Clinical aspects

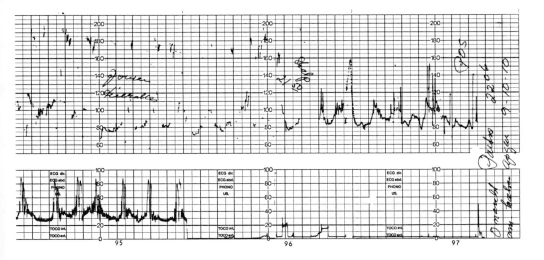

Fig. 12.1 A primipara at 38 weeks gestation in the second stage of labour. External (ultrasound) record of FHR in the first part of the trace showing signal loss and short episodes of doubling of the apparent FHR — from 90 (true rate) to 180 bpm. A fetal scalp electrode was attached at mark 96 and gave a good record. The baseline FHR was between 80 and 90 bpm with low variability. Apgar scores were normal.

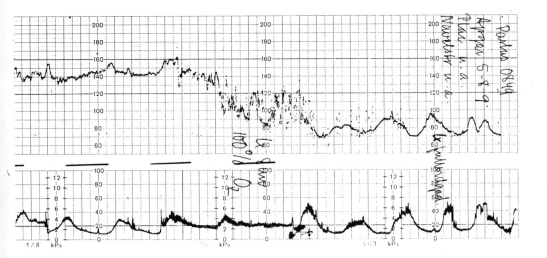

Fig. 12.2 The second stage of labour in a multipara showing a normal FHR with low variability followed by a progressive bradycardia to below 90 bpm, initially with increased variability but then with loss of variability. Apgar scores were 5, 8 and 9 at 1, 5, and 10 min. Delivery should be completed within 20 min of the onset of a bradycardia such as this.

a good correlation with the 5-min Apgar score and neonatal complications secondary to hypoxia is only seen when the cord artery pH is below 7.

Interpretation of the FHR pattern

Problems with interpretation of second stage FHR patterns are common. Although the autocorrelation technique has improved the quality of signals, external records may still give tracings of poor quality (Fig. 12.1). All parameters of the FHR pattern are more difficult to assess when maternal activity, pushing efforts, and maternal breathing all influence the quality of the CTG.

Baseline variability is difficult to assess unless very low (FIg. 12.2). Although decelerations may be difficult to classify it is not usually helpful to do so. Research concerning second stage patterns is scanty but it is possible to conclude a few guidelines from the available literature.

Classification

Several typical FHR patterns have been described:

1. Type 1 (10–15 per cent): normal with frequent accelerations (Figs 12.3 and 12.4).

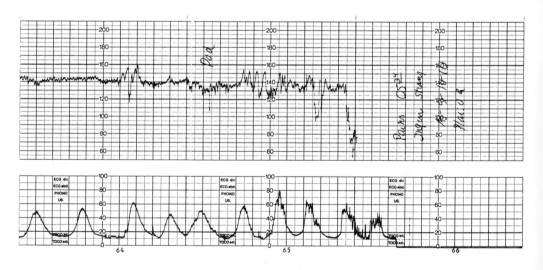

Fig. 12.3 A normal FHR pattern in a short second stage.

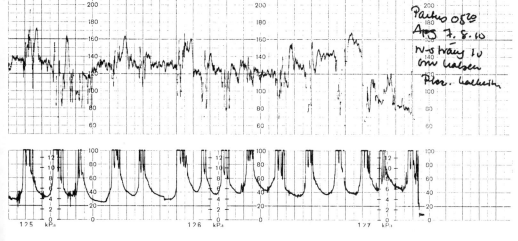

Fig. 12.4 A normal FHR pattern during pushing in the second stage.

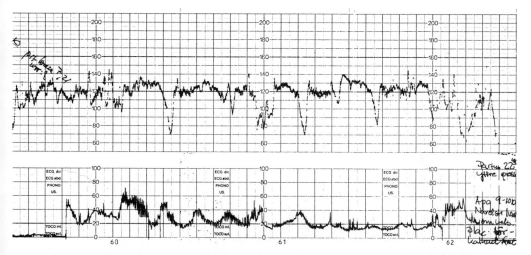

Fig. 12.5 Occasional decelerations during the second stage with a normal baseline FHR and variability. Scalp blood pH was 7.21 at the beginning at the record. Apgar scores were normal. The umbilical cord was looped once around the fetal neck.

2. Type 2 (35–40 per cent): normal baseline FHR and normal or high variability. Decelerations of predominantly variable type with most contractions. Combined decelerations can also appear (Figs 12.5 and 12.6).

3. Type 3 (10–15 per cent): sustained FHR tachycardia or episodes of tachycardia after decelerations. Low or absent baseline variability although high variability can also be seen. Pronounced decelerations with longer duration than in previous type 2 pattern (Figs 12.7 and 12.8).

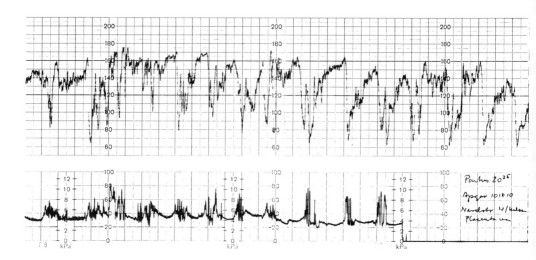

Fig. 12.6 A second stage FHR of between 135 and 155 bpm and high variability, especially during the variable and combined decelerations. Apgar scores were normal. The cord was looped once around the fetal neck.

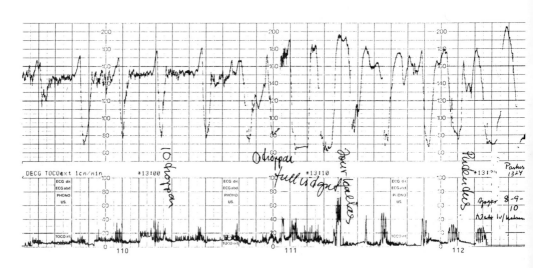

Fig. 12.7 Initially, this record shows a pattern of variable decelerations with accelerations and normal baseline variability. There is an abrupt change in the baseline to become a tachycardia with loss of variability at the onset of the second stage. The second stage was short and Apgar scores were normal.

Interpretation of the FHR pattern

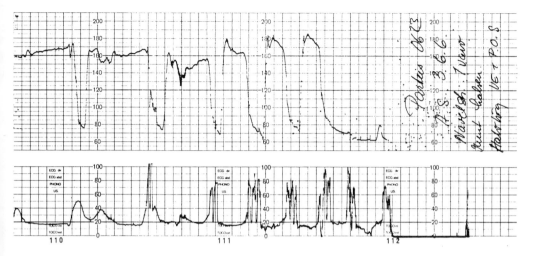

Fig. 12.8 Transition to the second stage was accompanied by loss of variability, which was already low, and a further rise in FHR. The variable decelerations became more prolonged. A bradycardia developed as the fetal head crowned and so delivery of the fetus was expedited by ventouse extraction. Apgar scores were 3, 6, and 6 at 1, 5, and 10 min. The cord was looped once around the fetal neck.

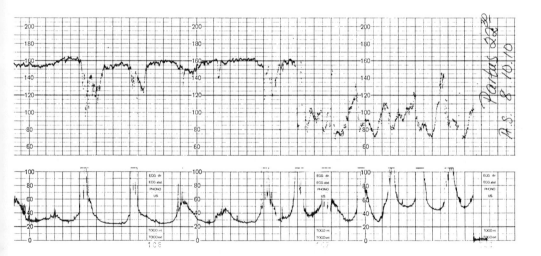

Fig. 12.9 A bradycardia occurred for the last 10 min of this short second stage of labour. The FHR rose between contractions but the frequency of contractions prevented a return to the normal FHR. Apgar scores were normal.

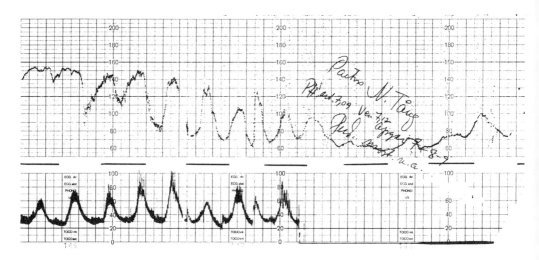

Fig. 12.10 A second stage complicated by late decelerations and progressive bradycardia. The FHR finally remained below 90 bpm with loss of variability. A mid-cavity forceps delivery was performed within 10 min of the onset of bradycardia (at the end of the panel). Apgar scores were 7 and 8. Cord arterial pH was 7.09 and venous pH was 7.17.

4. Type 4 (25–30 per cent): FHR bradycardia (<100 bpm). This pattern can follow other types, usually type 3. In a mild form the FHR picks up between contractions, reaching above the 100 bpm level. A more serious variant is a progressive bradycardia with gradual reduction in FHR with each push. Loss of variability occurs, especially with sustained bradycardia below 80 bpm. The assessment of variability is of value since fetal condition rapidly deteriorates in association with loss of variability (Figs 12.9 and 12.10).

Correlation between these types of FHR patterns with Apgar scores or cord arterial pH may be poor, as many other factors influence the fetal condition at birth. Types 1 and 2 are often considered normal second stage patterns and type 3 an intermediate form. Cases with type 4, particularly when bradycardia and loss of variability prevails, merit intervention if persisting beyond 10 min (see Fig. 12.2).

Studies correlating the length of the second stage to different FHR patterns and the condition of the newborn have suggested that type 4 often occurs after 30 min pushing. Severe bradycardia is an important finding during the second stage. With a monitor, in contrast to intermittent auscultation, the correct heart rate will be recorded and, most important, an assessment of variability will be possible (Figs 12.11 and 12.12). Classification of different decelerations seems to be of limited value.

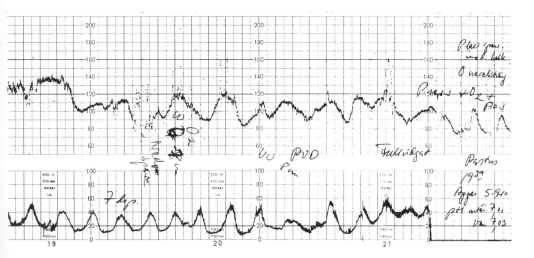

Fig. 12.11 A multipara at 43 weeks with slow progress in labour had a normal FHR pattern preceding this panel. A bradycardia of between 90 and 120 bpm occurred suddenly with reduced variability. Delivery occurred 25 min later and Apgar scores were 5 and 9. Cord arterial pH was 7.01 and venous pH was 7.03.

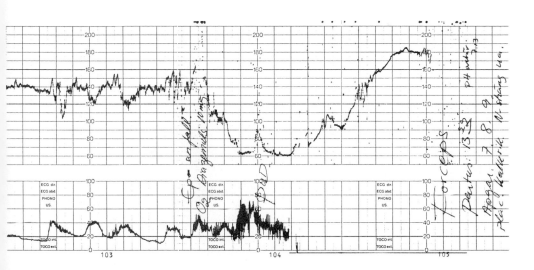

Fig. 12.12 A primipara who had an epileptic fit during the second stage after a normal first stage. Diazepam 10 mg i.v. (4 min before mark 104) was followed by a bradycardia with loss of variability which then changed to a tachycardia. Delivery was expedited by forceps. Apgar scores were normal. Umbilical artery pH was 7.13.

Summary

Use of FHR monitors in the second stage of labour is advised when this stage is prolonged, when there is breech presentation, a twin delivery, an instrumental delivery, or when other signs of fetal distress appear, such as meconium-stained liquor or auscultated FHR abnormalities.

13
Preterm fetal heart rate patterns

Introduction

Cooperation between obstetrician and neonatologist has resulted in the development of a combined approach to the management of preterm birth. After the introduction of continuous FHR monitoring, which was initially used in term labour after a complicated pregnancy, increasing experience of its use during preterm labour has contributed to our knowledge of FHR control at this time.

Clinical aspects

Increasing use of continuous FHR monitoring during preterm labour also reflects obstetric willingness to perform a Caesarean section for a fetal indication during the preterm period. The Caesarean section rate for preterm babies (birthweight less than 1500 g) rose from 10 per cent in 1973 to about 50 per cent in 1984 in Sweden. This rise was the result of a change in attitude towards operative delivery for the small fetus, which, although associated with a continuing fall in perinatal mortality rate from 60 to 25 per cent, was not the only factor contributing to the improved outcome of these cases. During the same period of time the neonatal survival rate in Sweden for newborns with a birthweight of between 1000 and 1500 g increased from 50 to 75 per cent, and from 22 to 50 per cent of those with a birthweight less than 1000 g.

Electronic monitoring is not without problems in the preterm period. Technical difficulties and interpretation problems are common before 30 weeks. It is often necessary to use external ultrasound for fetal monitoring and the small size of the fetus increases the chance of insonating maternal vessels and producing a maternal pulse rate. Spontaneous, uncomplicated, preterm deliveries are not frequent and labour is usually rapid. Assessment of fetal acid–base status is not possible with intact membranes.

Factors influencing the FHR

There are no established criteria for the interpretation of the FHR before 30–32 weeks. It is important to consider the effects of gestational age but there is a paucity of knowledge about such tracings.

Antenatal CTG

FHR falls as gestation progresses. The major part of this occurs before 30 weeks and, thereafter, the fall in rate is slight—noted primarily during episodes of fetal quiescence. The reduction is believed to reflect developing parasympathetic (vagal) influence on the fetal heart, which also results in changes in short-term and long-term variability during pregnancy. FHR variability increases with increasing gestation and depends upon the behavioural state of the fetus. Fetal activity is associated with increased/high variability whilst fetal quiescence is associated with reduced/low variability. The influence of the fetal rest–activity cycle only becomes apparent during the third trimester and true behavioural states do not exist until 36 weeks.

Fetal activity at around 28 weeks is different to that at 38 weeks. Preterm fetuses have short episodes of activity and quiescence (about 10 min each) and change from one to the other about six times each hour. Between 38 and 40 weeks, these episodes last longer and the fetus changes state only once or twice each hour. Quiet periods are shorter than active periods but most are shorter than 40 min.

The presence of accelerations is the key to a reactive pattern on an antepartum CTG (non-stress test). The number of fetal movements is greater in the preterm period but they are less accentuated, less powerful, and of shorter duration than later in gestation. The number of accelerations increases towards the end of the pregnancy, with the greatest increase occurring between 28 and 34 weeks. The FHR response to movements is of special interest. Studies have shown that the fetal heart responds differently to movements in the early preterm period compared with term pregnancy (Table 13.1).

Between 25 and 30 weeks gestation, decelerations are more common than accelerations in response to fetal movements. Most of these decelerations are of short duration (15–30 s) with an amplitude of 15–30 bpm (Figs 13.1, 13.2 and 13.3). After 30 weeks, accelerations are more common than decelerations in response to fetal movements (Figs 13.4 and 13.5). At term, decelerations are not seen with fetal activity. A comparison between typical normal FHR records at 30 weeks and 40 weeks is shown in Figure 13.6.

Table 13.1 Fetal heart rate responses to movements at different gestations

	25–30 weeks (%) ($n=114$)	31–35 weeks (%) ($n=229$)	36–40 weeks (%) ($n=190$)
Acceleration	17.5	54.8	59.0
Acceleration + deceleration	3.5	32.2	35.3
Deceleration	35.1	3.5	–

Factors influencing the FHR

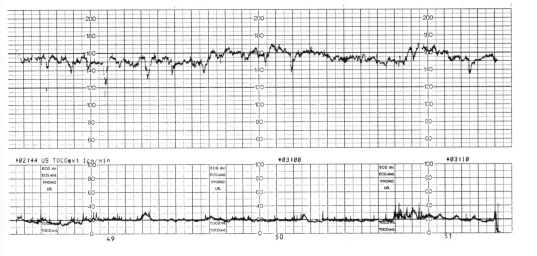

Fig. 13.1 An antenatal CTG at 26 weeks. The baseline is between 150 and 160 bpm with low variability. Small, 'spike' decelerations occur in association with fetal movements. A few small accelerations also occur. This is a normal FHR appearance during the second trimester.

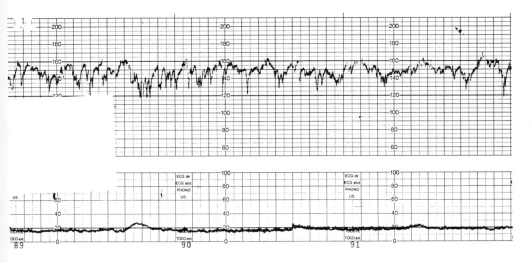

Fig. 13.2 An antenatal CTG at 27 weeks after prelabour rupture of the membranes. The baseline rate is between 150 and 155 bpm with normal variability and frequent small decelerations of no pathological significance.

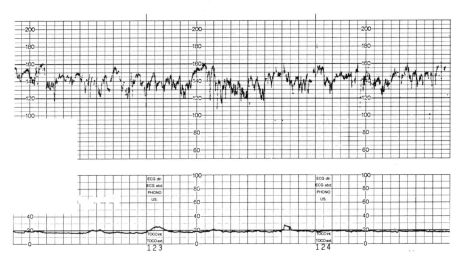

Fig. 13.3 An antenatal CTG at 27 weeks. The baseline FHR is 140 bpm with normal variability and frequent accelerations.

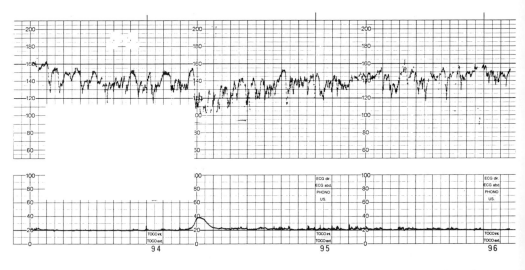

Fig. 13.4 An antenatal CTG at 30 weeks. Movements are associated with combinations of accelerations and 'spike' decelerations.

Interpretation of the preterm FHR is difficult if records are non-reactive with reduced variability. Decelerations occur with growth-retarded fetuses and oligohydramnios (Fig. 13.7). Although a similar mechanism may operate to produce small (type 0) dips (with a duration less than 15 s and an amplitude

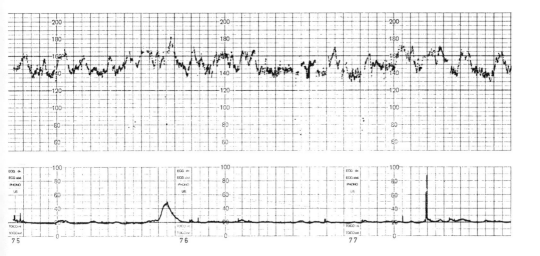

Fig. 13.5 An antenatal CTG at 33 weeks. The baseline FHR is 140 bpm with normal variability and reactivity. Frequent accelerations are seen with fetal movements.

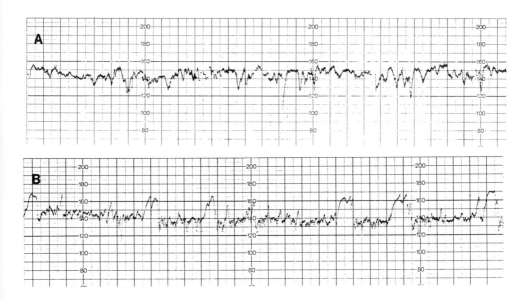

Fig. 13.6 A, A typical, normal, antenatal CTG at 30 weeks with small decelerations; B, the same case 10 weeks later shows a normal pattern with accelerations.

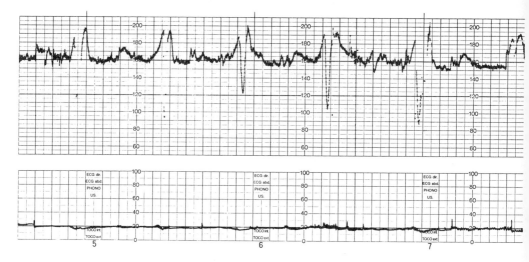

Fig. 13.7 An antenatal CTG at 32 weeks with prelabour rupture of the membranes and oligohydramnios. The baseline FHR is 160 bpm with normal variability. Recurrent short variable decelerations appear to interrupt FHR accelerations.

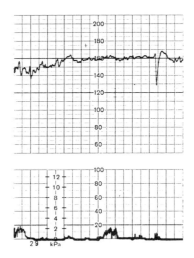

Fig. 13.8 Fetal activity with a 'spike' deceleration (also termed dip 0) at the end of this preterm record.

between 15 and 30 bpm) by cord disturbance before 30 weeks, this normal feature of the preterm FHR may indicate immature control of the cardiovascular system at that time (Fig. 13.8). Such short, sharp, decelerations can be seen later in pregnancy in association with fetal hiccups.

A normal, reactive, pattern is often observed in the early preterm period. Interpretation of these records is simple, as is the case with severe and ominous FHR changes, such as loss of variability and shallow decelerations. Repetitive late decelerations in association with spontaneous uterine activity are a reliable indication of placental insufficiency (Figs 13.9 and 13.10).

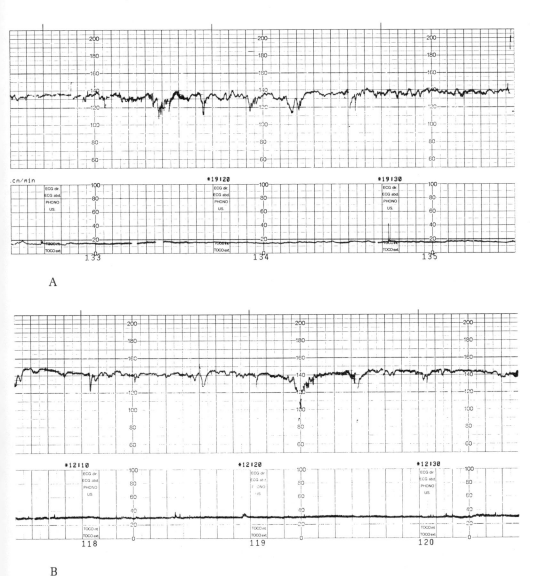

Fig. 13.9 (caption overleaf).

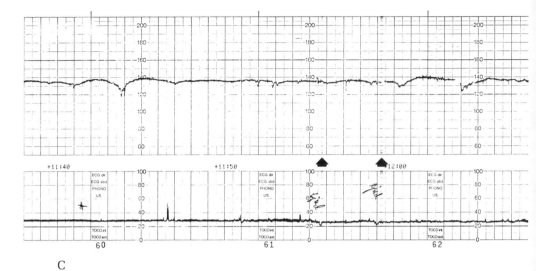

C

Fig. 13.9 A primipara with severe pre-eclampsia at 27 weeks gestation. A, A normal CTG for the gestation; B, 2 days later, there is a slight reduction in variability; C, another 2 days later, the baseline FHR remains unchanged at 135 bpm but the variability is now definitely reduced. Sporadic shallow decelerations have replaced the 'normal' spike type 0 dips. Vibroacoustic stimulation applied twice (at the arrows) did not produce a response in the fetus. A Caesarean section was performed that day. Apgar scores were 2, 6, and 8 at 1, 5, and 10 min, the birthweight was 770 g.

Preterm labour

Experience has increased our knowledge about FHR changes and fetal acid–base balance during preterm labour. Different abnormalities occur in preterm labour, especially prior to 34 weeks, compared with after 36 weeks. Table 13.2 shows the changes in baseline, variability, and decelerations at these times.

Tachycardia and reduced variability are common in labours before 34 weeks without associated fetal acidosis. Variable decelerations occur in 70 per cent of preterm fetuses between 28 and 30 weeks and in more than 50 per cent in the more mature group of fetuses. Tachycardia, reduced variability, and variable decelerations seem to reflect fetal immaturity.

Reduced or absent variability, late decelerations, and complicated variable decelerations are frequently associated with fetal hypoxaemia or even acidosis in pregnancies at risk of placental insufficiency. Fetal acidosis can appear abruptly during preterm labour (Table 13.3). It is possible that the capacity of the preterm fetus to handle stress during labour is reduced, to some extent, compared with the term fetus.

14
Ante

Introduction

The antepartum CTG, or non-stress test (NST), is the most common method of assessing fetal well being in the second half of pregnancy. Regular contractions are usually absent and so the FHR will not be subject to any interference with placental function during uterine activity. This contrasts with the oxytocin challenge test (OCT), described later in this chapter (see p. 250), where contractions are stimulated in order to assess placental function by the effect on the FHR.

Descriptive features

The following features are described when interpreting the antepartum CTG: (i) sporadic and periodic accelerations; (ii) fetal reactivity; (iii) baseline heart rate; (iv) baseline variability; and (v) decelerations. A record with at least two sporadic FHR accelerations and a normal baseline rate, without decelerations,

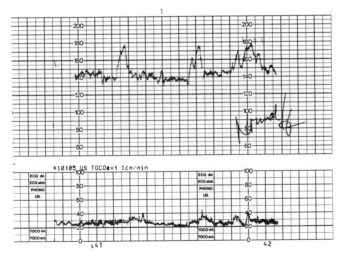

Fig. 14.1 A reactive antepartum CTG, also called a non-stress test (NST).

is described as normal and reactive (Fig. 14.1). It is accepted that these features indicate fetal health and, therefore, prolonged records are not necessary, especially during third trimester. The risk of intrauterine death is negligible during the 24 hours after such a record unless due to an unpredicted problem (see Fig. 15.2). The fetal death rate is about 3 per 1000 over the following week after a normal record in a high risk pregnancy.

Accelerations (fetal reactivity)

There is no absolute agreement on interpretation of the antepartum CTG, and interobserver variation is significant. However, it is generally agreed that fetal reactivity, the presence of accelerations with fetal movements, is of importance for the recognition of fetal well-being. Fetal reactivity indicates integration between the brain stem control mechanisms for body movement and heart rate, which is lost when the fetus is significantly hypoxic. Baseline variability may be difficult to assess if monitored with external ultrasound but accelerations reflect normal variability. Accordingly, absence of accelerations may occur with central nervous depressant drugs, quiet fetal behavioural state, and congenital anomalies, and these need to be excluded before fetal hypoxia can be suspected.

Accelerations greater than 10 bpm occur with 99 per cent of fetal movements lasting more than 3 s in fetuses between 28 and 42 weeks. An average lag time of 1.3 s exists between the start of a movement and the start of acceleration. Practically all accelerations of more than 20 bpm are associated with fetal movements. This contrasts with the decelerative pattern seen between 20 and 30 weeks (see Chapter 13).

The number of accelerations needed to classify a CTG as reactive remains a subject of debate in the literature. If, as some have suggested, five accelerations are required in a 20-min period, up to 35 per cent of CTG records will be non-reactive and prolongation of the record will be necessary. Fetal condition is satisfactory in 98.5 per cent of such cases, even though five accelerations may only be recorded in an 80-min period.

Most accelerations accompany powerful fetal movements. When monitoring fetal activity continuously by ultrasonic scanning it has been demonstrated that 75 min may pass without fetal activity. No accelerations occur during such periods of fetal quiescence. The consequence of such fetal behaviour—prolongation of the CTG—is necessary until fetal activity occurs. To save time it has been suggested that the fetus be stimulated, either by manual manipulation or vibroacoustic sound. Some have formalized this into a protocol such that stimulation is applied to provoke fetal activity if the CTG remains non-reactive after 20 min. We have found that, after this period of time, nine out of ten fetuses with a non-reactive Admission Test demonstrated reactivity after vibroacoustic stimulation (see Chapter 17).

Descriptive features

The baseline heart rate

Tachycardia

In association with fetal activity, transient episodes of tachycardia (above 150 bpm) may occur (Fig. 14.2). Other causes of tachycardia at term include maternal fever, maternal thyrotoxicosis, or maternal stress with increased

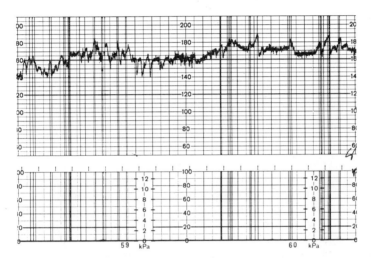

Fig. 14.2 A reactive antepartum CTG with tachycardia due to fetal activity. The vertical lines of the event marker indicate the movements perceived by the mother.

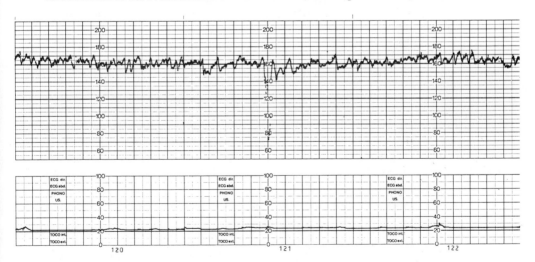

Fig. 14.3 An antepartum CTG showing a FHR tachycardia with normal variability but no accelerations. This is a suspicious (non-reassuring) record.

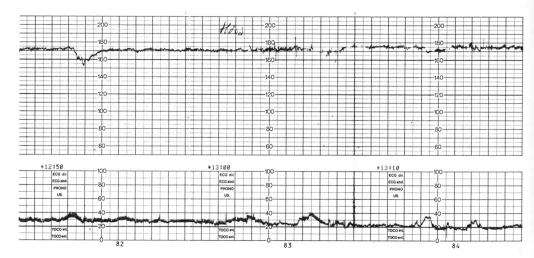

Fig. 14.4 A pathological CTG at 38 weeks of pregnancy complicated by intrauterine growth retardation. The baseline FHR is 170 bpm with low variability and no accelerations. A late decelerations is seen with one of the uterine tightenings.

sympathetic tone (anxiety, agony, pain). β-mimetic drugs and atropine may also induce maternal and fetal tachycardia. In the absence of an explicable cause, a CTG showing a tachycardia should be repeated within 24 h (Fig. 14.3). Tachycardia with loss or absence of FHR variability may be a sign of fetal hypoxia (Fig. 14.4). Before intervention is considered it might be advisable to exclude a fetal malformation, either by ultrasound scan or by fetal blood sampling (cordocentesis) for chromosomal abnormalities.

Bradycardia

Bradycardia—a baseline heart rate of less than 120 bpm—is uncommon during pregnancy (Figs 14.5 and 14.6). Records should not be taken with the mother in the supine position because of the risk of hypotension causing fetal bradycardia. Cardioselective β-blockers for the treatment of hypertension may also lower the baseline FHR by as much as 10 bpm. A sustained, pronounced bradycardia, with total loss of variability, may be seen in cases of atrioventricular block (see Fig. 6.29) and bradycardia may be a terminal pattern, particularly after placental abruption (see Chapter 15).

It is important to exclude accidental identification of maternal heart rate if the CTG shows a bradycardia. This is especially true during pregnancy, when β-mimetic drugs to inhibit preterm labour are administered, producing a mild maternal tachycardia of between 100 and 110 bpm (see Fig. 2.9).

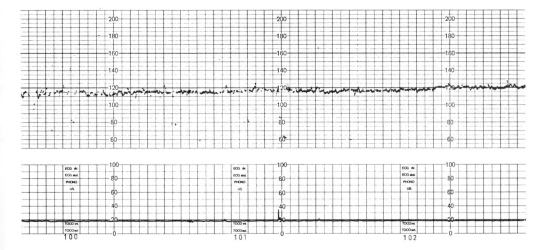

Fig. 14.5 A pregnancy complicated by gestational diabetes at 34 weeks with fetal growth retardation. The baseline FHR is 115 bpm with low variability and with no accelerations. This is a pathological test result.

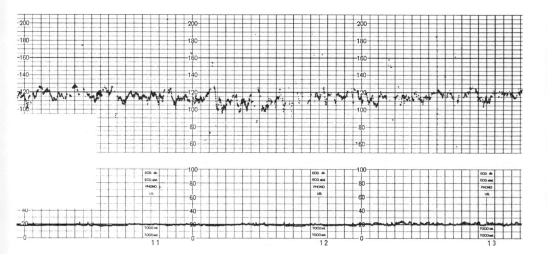

Fig. 14.6 A pregnancy complicated by pre-eclampsia at 37 weeks. The baseline FHR is 115 bpm with normal variability. This is a suspicious CTG in the absence of accelerations.

Baseline variability

Short-term variability is difficult to evaluate visually unless the baseline is sharply reduced. Long-term variability is more easily estimated by measuring the bandwidth or amplitude of the oscillations (see Chapter 5). Long-term variability may also be assessed by counting the number of times the FHR trace crosses the baseline (oscillation frequency). Although a laborious way of estimating long-term variability, it is included in some of the CTG scoring systems.

From animal experiments it is known that increased variability occurs with acute fetal hypoxia. However, it is not known whether this is true for the human fetus. Another association with this pattern (sometimes in conjunction with variable decelerations) is oligohydramnios, which predisposes the umbilical cord to intermittent compression during fetal activity. The most common reason for increased variability is, however, an active fetus.

Reduced variability, sometimes with a mild tachycardia, occurs with chronic fetal hypoxia. This pattern may be seen in cases with fetal growth retardation associated with chronic placental insufficiency.

Preterm fetuses before 30 weeks may show reduced variability. Central nervous system depressant drugs given to the mother will also reduce variability. It is possible that cardioselective β-blockers (for treatment of maternal hypertension) also reduce variability. Finally, consideration of fetal behavioural states and gestational age are prerequisites for correct interpretation (Fig. 14.7).

Decelerations

Decelerations occur with fetal activity or contractions. Decelerations that are late or variable (especially if pronounced) with uterine activity (Fig. 14.8) are

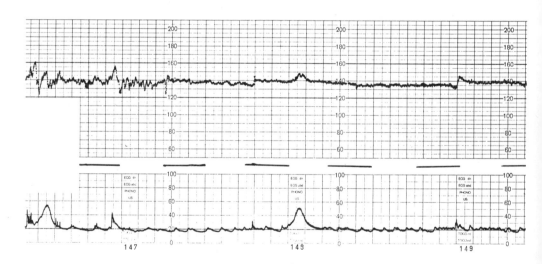

Fig. 14.7 A reactive CTG in a pregnancy at 39 weeks with pregnancy-induced hypertension. Accelerations are seen in the first part, which is followed by a quiet fetal state with low variability. Prolongation of test will give no further information.

Descriptive features 229

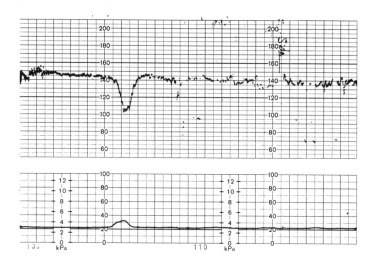

Fig. 14.8 An antenatal CTG at term. The baseline FHR is 145 bpm with low variability. There is a variable deceleration with a uterine tightening. The CTG was reactive the next day in the absence of uterine activity.

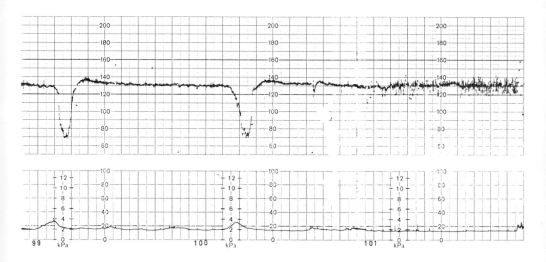

Fig. 14.9 A pregnancy at 37 weeks complicated by intrauterine growth retardation. Braxton Hicks tightenings are associated with pronounced late decelerations. The baseline FHR is 130 bpm with absent variability (note the artefactual jitter at the end of the record). There is a slight tendency for the FHR to overshoot after the decelerations. This is a pathological (terminal) FHR pattern. An emergency Caesarean section was performed after this test and a small-for-gestational age baby in good condition was delivered.

230   Antenatal cardiotocograph

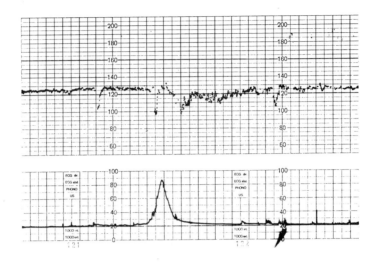

Fig. 14.10 An antenatal CTG showing mild variable decelerations with fetal movements (seen as slight deviations on the toco trace).

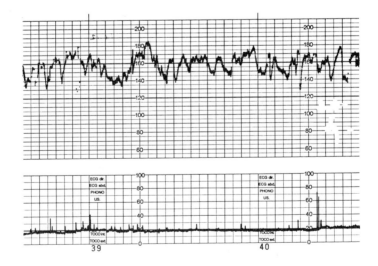

Fig. 14.11 An antenatal CTG at 29 weeks showing frequent variable decelerations in an active fetus.

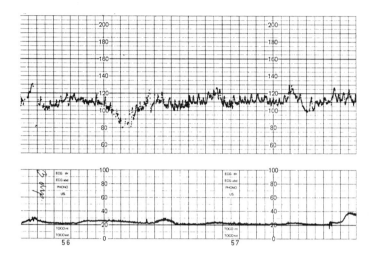

Fig. 14.12 A transient fall in FHR unrelated to uterine activity. There is a baseline bradycardia but accelerations are present. This may have been the result of supine maternal position with aortocaval compression.

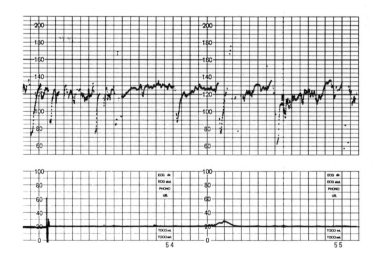

A

Fig. 14.13 (caption overleaf).

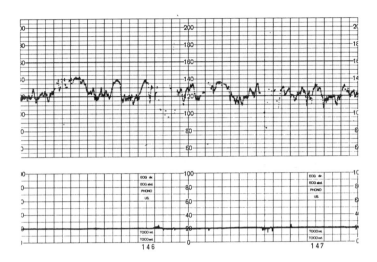

Fig. 14.13 Antepartum CTG records from a growth retarded fetus at 33 weeks. A, Frequent variable decelerations; B, a normal, reactive CTG the next day.

important signs of reduced placental function and failure of fetal adaptation. Such appearances have been described as a terminal pattern (Fig. 14.9). The biophysical profile may give additional, important information about fetal breathing movements, fetal activity, and the amount of liquor.

Decelerations associated with fetal activity are usually variable, although they may be uniform (Figs 14.10, 14.11, 14.12 and 14.13). Decelerations of this type have been observed in up to 10 per cent of recordings, and are similar to the variable decelerations in labour. Variable decelerations on antepartum CTG records indicate the need for monitoring in labour in order to detect possible cord compression.

The umbilical cord may be compressed by the fetus or between the fetus and uterine wall, particularly if there is oligohydramnios. Variable decelerations are not uncommon in association with contractions and fetal activity in such cases, particularly with a growth retarded fetus. As previously mentioned, a decelerative pattern is common before 30 weeks.

Recording procedure and interpretation

Standardization of performance and interpretation of the CTG is necessary to optimize results. The following procedural standards have been recommended (modified after Lavery 1982). The test should be performed after a meal, when fetal activity is more likely. The mother should lie in the lateral position and should be asked to note fetal movements. The recording time should be at least 20 min. If the CTG remains non-reactive the recording should be prolonged or stimulation (such as vibroacoustic stimulation) applied.

Interpretation

The following criteria for interpretation of the CTG have been suggested (Evertson *et al.* 1979). A reactive record requires two accelerations of at least 15 bpm lasting at least 15 s in association with fetal movements. A non-reactive record has one or fewer accelerations in two 20-min periods, the second after fetal stimulation. Some advocate an OCT after a non-reactive NST.

Such guidelines for interpretation of the CTG may create problems. As previously stated, a term fetus spends about one-third of its time in a quiet, inactive, state, during which it can hardly be expected to produce a reactive CTG. This classification does not consider baseline variability nor decelerations. Another problem is the high number of CTG tests that would require an OCT due to absence of reactivity. Published reports using the above protocol indicate that an OCT may be required in up to 35 per cent of cases.

European practice does not favour the OCT and prefers a greater degree of analysis of the FHR pattern. The first step after a non-reactive test is to prolong the recording, up to 2 h if necessary (fetal activity may be absent in healthy fetuses for up to 90 min). It has been shown that accelerations may be absent for more than 64 min in 30–40 per cent of all records at term.

Classification of the CTG

Different scoring systems have been described to consider different parameters of the antepartum CTG. Some ascribe points to each parameter, rather like the Apgar score. An example of this is the Fischer score, which assesses five parameters and has a maximum possible score of 10 (Table 14.1). The record is made for 30 min. Variability amplitude corresponds to band-width and the

Table 14.1 The Fischer score for the antepartum CTG

Parameter/score	0	1	2
Baseline (bpm)	<100 >180	100–120 160–180	120–160
Amplitude (bpm)	<5	5–10, >30	10–30
Oscillations/min	<2	2–6	>6
Accelerations	None	Periodic	Sporadic
Decelerations	Late large variable	Variable	None dip 0

frequency is the number of times the baseline is crossed (long-term variability). A total sum of between eight and ten indicates good fetal condition, between five and seven is suspicious (OCT recommended), and less than five is ominous.

Solum and co-workers (1979, 1980) suggested classification into four groups (Table 14.2), which was slightly modified by Montan *et al.* (1985). In practice, between 85 and 90 per cent of all antepartum CTG records are normal (Fig. 14.14), between 6 and 8 per cent are suspicious (Fig. 14.15), and 1 to 2 per cent are pathological (Figs 14.16 and 14.17). Only 1 per cent of low-risk cases have CTG records in class 3 or 4 (Table 14.3).

Table 14.2 Classification of the antepartum CTG into four classes

Class	Baseline	Variability	Accelerations	Decelerations
1 Normal (Fig. 14.14)	120–160	>10 bpm	>2 in 20 min	None
2 Suspicious (Fig. 14.15)	100–120 or 160–180 bpm	>25 bpm 5–10 bpm	None	None
3 Mildly pathological (Fig. 14.16)	<100 bpm or >180 bpm	5–10 bpm		Moderate variable
4 Seriously pathological (Fig. 14.17)		<5 bpm		Pronounced variable or late

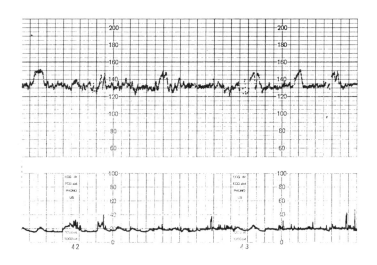

Fig. 14.14 A normal, reactive CTG.

Recording procedure and interpretation 235

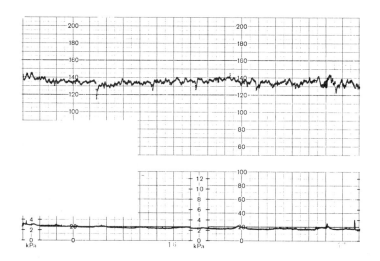

Fig. 14.15 A non-reactive, suspicious CTG.

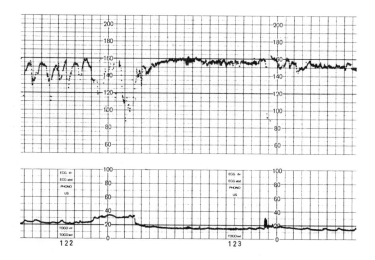

Fig. 14.16 A mildly pathological CTG.

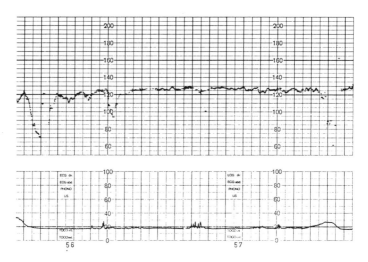

Fig. 14.17 A severely pathological, ominous CTG.

Table 14.3 Results of antepartum CTG testing

Class	Low risk (411 cases)	High risk (401 cases)
1 Normal	93.5	77.6
2 Suspicious	5.4	7.2
3 Mildly pathological	0.9	7.2
4 Seriously pathological	0.2	8.0

It is unclear whether scoring systems have improved results. Hage (1985) found a better kappa index for binary (0.60) than for a five-category classification and Flynn et al. (1982) found a better correlation with a binary system than with three semi-quantitative systems. On the other hand, Trimbos and Keirse (1978) considered a ten-point scoring system superior to a three-category system.

Regardless of the system of classification, there are many problems associated with interpretation of antepartum CTG records, even though there are fewer confounding factors compared with interpretation during labour. Inter- and intraobserver variation in interpretation is high. Borgatta et al. (1988) found that Cohen's kappa for interobserver variation was between 0.380 and 0.784 and, most disappointingly, that observers disagreed 28.4 per cent of the time. This variation was not related to age of observer, experience, or the number of tests. Some misinterpretations can be attributed to difficult records in the preterm period and the fact that fetal behavioural states are not considered.

Objective analysis of the antepartum CTG

Designed to overcome subjective variation in visual analysis of antepartum CTG records, the numerical analysis system of Dawes and Redman is now commercially available (System 8000, Oxford Sonicaid Ltd, Chichester, UK). Data reduction is achieved by deriving a mean pulse interval from the FHR every 3.75 s. This gives 16 data points per minute, from which the baseline is derived and plotted. Deviations (accelerations and decelerations) from the

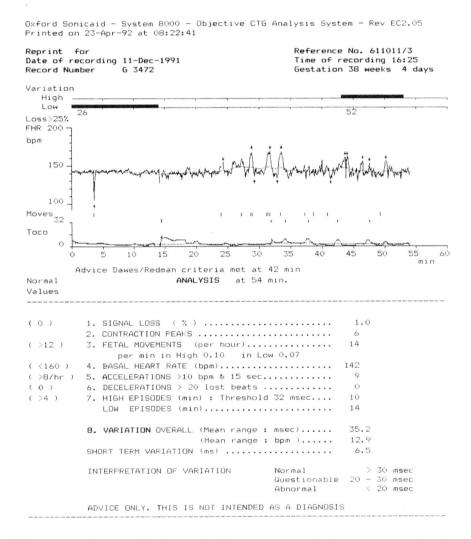

Fig. 14.18 A normal CTG record with numerical analysis, as presented by the System 8000 monitor (Oxford Sonicaid Ltd, Chichester, UK).

baseline can therefore be recognized objectively. Variability is derived from the variation in mean pulse intervals and quantified in milliseconds (Fig. 14.18). Experience at Oxford has resulted in a large data bank and results have been correlated with fetal outcome to allow normal ranges to be ascertained for each

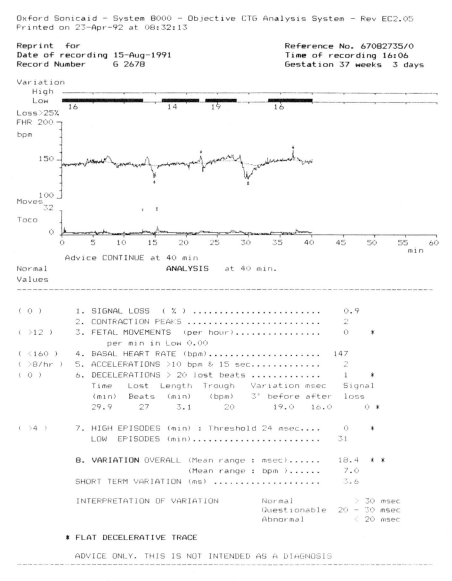

Fig. 14.19 A CTG at term in a small fetus with low variability, analysed by the System 8000 monitor.

derivative of the FHR. The programme is based in a personal computer, which receives the FHR and toco information from a standard antenatal monitor and a standard printer is required for the output of the numerical analysis.

Short-term variability of the FHR is now readily available by computer analysis of mean pulse intervals (Fig. 14.19). Recent data suggests that short-term

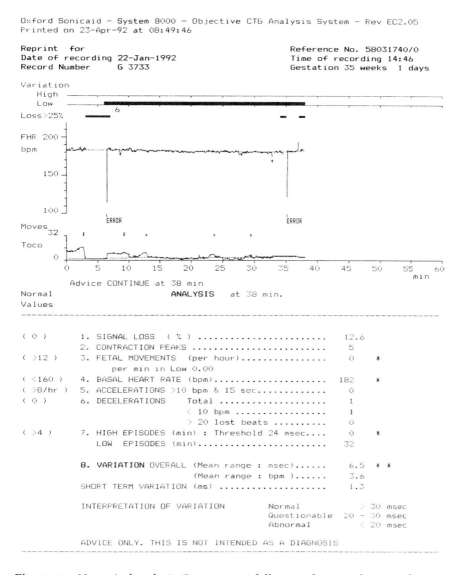

Fig. 14.20 Numerical analysis (System 8000) following abruptio placentae showing severely reduced overall and short-term variability (see Fig. 18.8).

variability is likely to be important for the recognition of chronic antepartum hypoxia secondary to placental insufficiency (Fig. 14.20).

The major advantage of such a system is the uniformity of analysis by the programme. System 8000 differs from other computer programs, which merely attempt to recognize the 'established' FHR patterns. Subjective differences in the interpretation of such patterns will not be overcome by uniform description of the CTG. However, whether numerical analysis will prove to be essential, or merely useful, in the context of clinical management, remains to be determined. Its potential for screening the pregnant population requires assessment.

Clinical use of antepartum FHR records (non-stress tests)

The CTG is widely used to assess fetal well-being during pregnancy. However, the method is not appropriate as a screening test for fetal compromise in normal pregnancies. Trimbos and Keirse (1978) performed 594 CTG records in 91 normal pregnancies between 34 and 40 weeks gestation and found no ominous patterns. However, 7.2 per cent were suspicious and at least one such record was seen in 37 per cent of all pregnancies. These results indicate the potential danger of false positive results in normal pregnancies if the method is used inappropriately. Testing normal pregnancies 'for reassurance' should be avoided.

In our opinion, an antepartum CTG should only be performed for a well defined indication. Routine testing in high-risk pregnancies has not been shown to be of value and the CTG should be regarded as a diagnostic rather than a screening test. Its probable value lies in its ability to indicate the success of fetal adaptation to acute and chronic placental dysfunction, examples of which are cases with suspected abruption of the placenta, oligohydramnios, and fetal growth retardation. In other clinical situations, such as hypertension, diabetes, and post-term pregnancy (>42 gestational weeks), use is common, although benefit from such testing, in the absence of other indicators of placental dysfunction (such as abnormal Doppler ultrasound and reduced liquor volume), is not clear.

Duration of FHR record

It is not necessary to prolong a CTG beyond the time taken to record reactivity. This may be as short as 10 min if two accelerations are present with normal baseline rate and variability. If non-reactive, the record should be continued for up to 2 h, unless stimulation produces a change of fetal state and a reactive CTG. The recording can be stopped as soon as FHR accelerations are seen.

Many non-reactive records considered suspicious after the first 30 min will be seen to be normal after prolonged testing. A vigorous fetus may even have a basal heart rate greater than 160 bpm associated with a transient period of continuous activity (awake state).

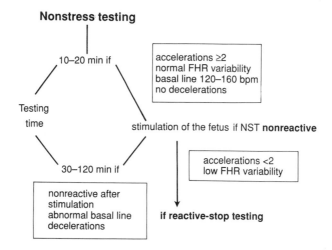

Fig. 14.21 A scheme for the use of antepartum cardiotocography.

With repetitive decelerations, a record up to 120 min is recommended to provide maximum information (Fig. 14.21). Prolonged records may also be indicated preterm or where the mother is on medication which may affect the FHR.

How often should the CTG be repeated?

The concept of routine weekly testing needs to be considered carefully to avoid unnecessary testing in women who do not really need FHR monitoring. It may be difficult to stop testing once it has been instituted. Repeated testing after a normal test is not usually indicated in low-risk women. The CTG should be repeated within 24 h if the first remains suspicious after prolonged testing. Management depends upon the clinical situation, and CTG tests should be performed whenever fetal condition requires instantaneous assessment.

Interpretation of an abnormal CTG result is also necessary in the light of the clinical circumstances. It should be remembered that the FHR indicates fetal condition at that moment and, broadly speaking, reflects fetal adaptation to possible placental insufficiency. Thus, successful adaptation to early stages of reduced placental function may mean that the CTG is still normal, giving no indication of the degree of placental dysfunction. An abnormal result indicates serious disturbance to fetal cardiovascular control, which may be a reflection of hypoxia and acidaemia. If the clinical circumstances support such an interpretation (maternal hypertension, fetal growth retardation, repeated placental abruptions) then appropriate clinical action may be considered. Other abnormalities that require appropriate management include oligohydramnios, congenital malformation, and rhesus isoimmunization. Ultrasound imaging and Doppler blood flow velocity waveforms may be useful in such cases. Cordocentesis may be appropriate if a chromosomal abnormality is suspected.

Frequent testing is indicated when the diagnosis is known and the condition of the fetus requires close monitoring. If the FHR is bradycardic, a fetal heart malformation should be considered and maternal autoimmune disease should be excluded. Of 12 cases with persisting FHR bradycardia (less than 100 bpm) a fetal heart malformation was diagnosed in four: isolated complete heart block was found in eight fetuses with normal heart anatomy and four of these mothers had a collagen disease. If the FHR appears sinusoid then rhesus isoimmunization, fetal bleeding or anaemia, or severe hypoxia should be considered.

A plan for continuous supervision of the fetus should always be made but protocols for testing intervals will depend upon the individual case. Repetitive late decelerations or pronounced variable decelerations are usually found in circumstances where intervention is considered appropriate. Such decelerations, together with loss of variability and an abnormal baseline rate, usually indicate the limit of fetal ability for adaptation to chronic placental dysfunction. The fetus is likely to be hypoxic and will be developing a metabolic acidaemia prior to asphyxial death; hence the description 'terminal pattern'. It is impossible to know exactly how serious the situation is for the fetus in such cases, and prediction of survival time is unreliable. Terminal patterns can continue for days but may also last for a few hours only (see Fig. 15.5).

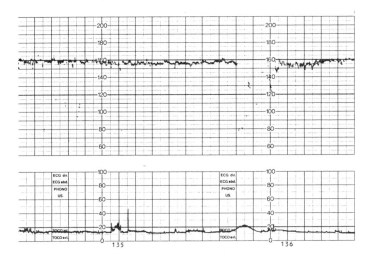

Fig. 14.22 A poor quality antepartum CTG with signal loss during a uterine tightening.

Technical problems

Poor technical quality of records may influence interpretation of the CTG. Even a short record may be enough to demonstrate accelerations and fetal reactivity. Baseline variability can not be interpreted if 'jitter' is present (see Chapter 2). Important information will be lost if the FHR record contains too much artefact or has a high rate of signal loss, particularly during fetal activity or uterine contractions, when a deceleration may be missed (Fig. 14.22). It may be necessary to hold the ultrasound transducer in position on the abdomen to get a good record.

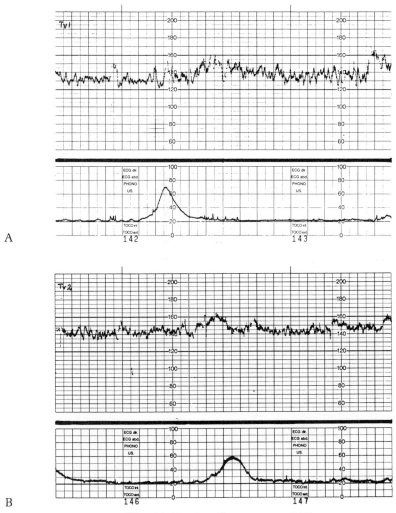

Fig. 14.23 (caption on page 246).

C

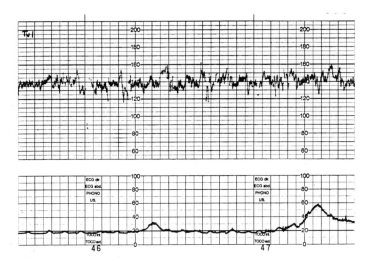

D

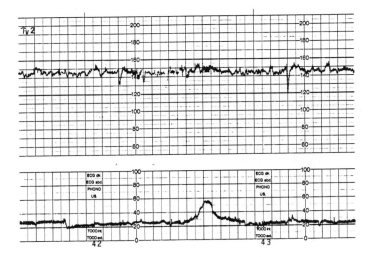

Fig. 14.23 (caption on page 246).

E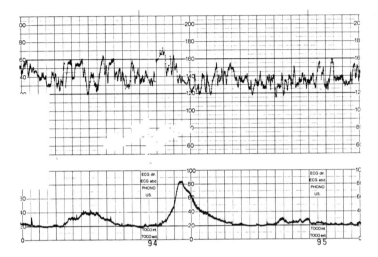

F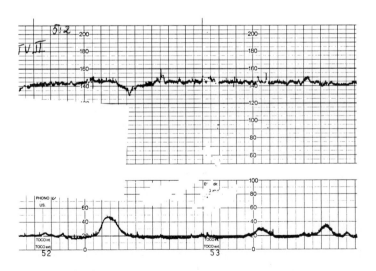

Fig. 14.23 (caption on page 246).

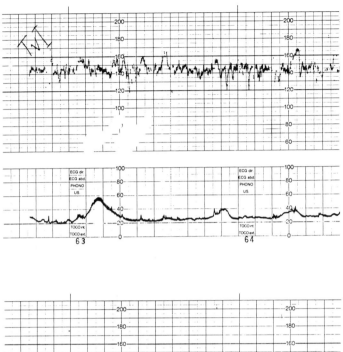

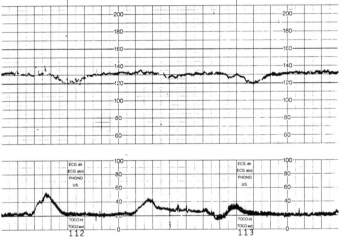

Fig. 14.23 CTG records from a primipara with a twin pregnancy complicated by hypertension. A and B, Normal, reactive, records at 34 weeks; C and D, 5 days later, the CTG of twin 2 (D) is non-reactive; E and F, 2 days later, the CTG of twin 1 remains normal whilst that of twin 2 is now severely pathological; G and H, 1 day later, twin 2 still has a severely pathological (terminal) CTG with absent variability and shallow decelerations. An emergency Caesarean section was performed that day. Apgar scores for twin 1 were normal and for twin 2 were 2, 5, and 9 at 1, 5, and 10 min. The birthweight of twin 1 was 2140 g and of twin 2 was 1210 g.

Abnormal uterine activity

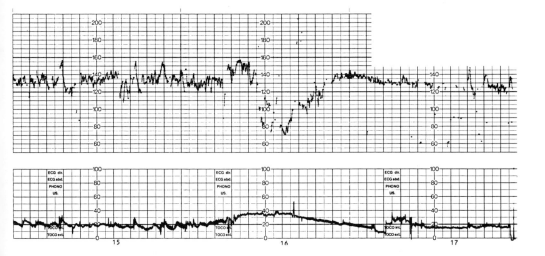

Fig. 14.24 An antenatal CTG at 39 weeks showing an abnormal uterine contraction (at mark 16), which produced a prolonged FHR deceleration. The CTG is reactive before and after the deceleration.

Special problems can be present in a twin pregnancy. One of the twins (often the second) may be difficult to monitor, especially preterm. It may also be difficult to establish that the origin of the obtained signal is the second twin. Such problems can be overcome by the use of two monitors to record the twins separately or by using a monitor with facilities for twin monitoring. A particular problem with twin pregnancy is the possibility that one may demonstrate FHR abnormalities whilst the other is normal (Fig. 14.23).

Abnormal uterine activity

Occasionally, women with normal pregnancies report a strong uterine tightening lasting several minutes. A prolonged deceleration may be observed with such a contraction and this is probably a normal fetal response rather than a sign of hypoxia. These decelerations look alarming because they are pronounced and of long duration (Figs 14.24 and 14.25). Maternal vena cava compression should be excluded.

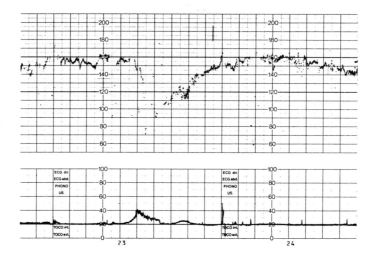

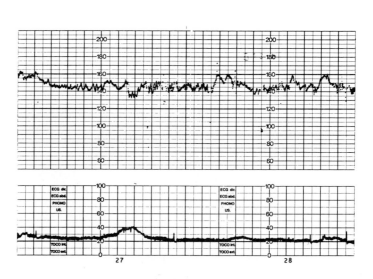

Fig. 14.25 An antepartum CTG at 37 weeks. A, Tachycardia with a pronounced deceleration following a uterine tightening; B, part of a prolonged record for 1 h, which showed a normal, reactive CTG 30 min later.

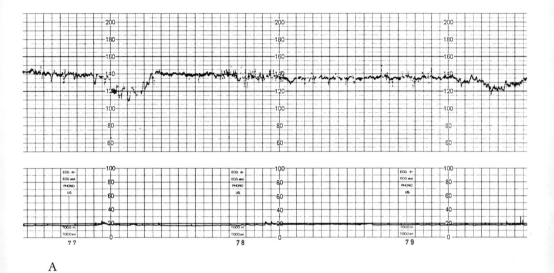

A

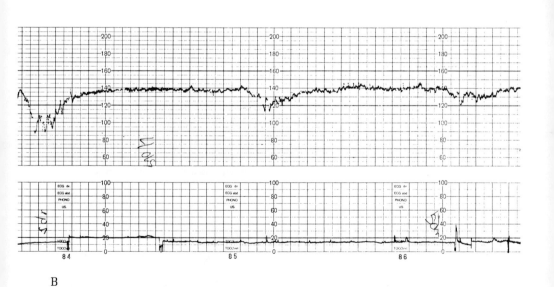

B

Fig. 14.26 (caption overleaf).

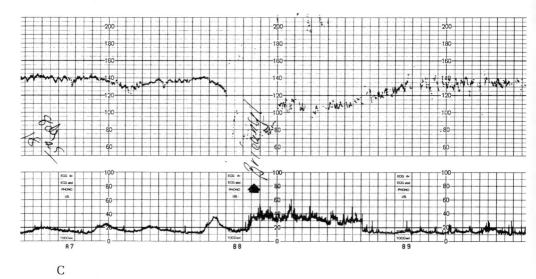

C

Fig. 14.26 Intrauterine growth retardation at 35 weeks. A, A possible late deceleration with low variability; B, an OCT was performed. Oxytocin infusion at 4 mU/min. Possible late decelerations but the poor contraction record makes interpretation difficult; C, infusion with 8 mU/min provoked overstimulation with a FHR bradycardia. Terbutaline (0.25 mg) i.v. relaxed the uterus. A Caesarean section was performed the same day. Apgar scores were normal. The umbilical arterial pH was 7.29 and venous pH was 7.32. The birthweight was 1540 g.

The oxytocin challenge test (OCT)

This is also called the oxytocin stress test and was introduced in the United States in 1972. It has been extensively used to assess placental function. The principle of the OCT is that an adapted, chronically hypoxic fetus, which may have a normal FHR pattern on the CTG (or non-stress test), will show FHR decelerations during uterine contractions. Contractions are induced by an intravenous infusion of oxytocin to test the fetal response to interruptions of the uteroplacental circulation. A fetus with reduced reserve is likely to respond with late decelerations. This is also the rationale of the Admission Test in early labour where the fetal response to spontaneous uterine activity is assessed.

Bradycardia is a common fetal response to acute interference of the placental circulation, such as cord prolapse or placental abruption. With chronic placental dysfunction and chronic fetal hypoxia, the CTG may not be abnormal in the absence of contractions. Eventually, baseline FHR and variability may be affected and late decelerations are the usual response to contractions.

The oxytocin challenge test (OCT)

Procedure

With the patient in a lateral position, the blood pressure should be checked. A CTG is obtained for 15 to 30 min prior to commencing oxytocin to record the possible presence of spontaneous contractions. An oxytocin infusion at a rate of 0.5 to 1 mU/min is then started and the dose is doubled every 15 min until contractions (with a duration of between 40 and 60 s and a frequency of three in 10 min) have been obtained. An infusion rate of 15 to 20 mU/min may be needed.

The infusion is stopped immediately if decelerations occur and the record is continued until the FHR pattern returns to normal again. A bolus injection of a β-mimetic drug may be considered if hyperstimulation produces prolonged decelerations or severe bradycardia (Fig. 14.26).

Interpretation

The OCT is considered negative if the FHR remains normal with three contractions in 10 min (Fig. 14.27). Late decelerations with less than half of the contractions are suspicious and a positive OCT is one where repetitive late or pronounced variable decelerations occur with most contractions (Figs 14.28 and 14.29).

A clear answer may not be obtained if hyperactivity (more than five contractions in 10 min) produces prolonged decelerations (longer than 90 s) or if there are technical problems in recording the FHR or obtaining sufficient uterine activity.

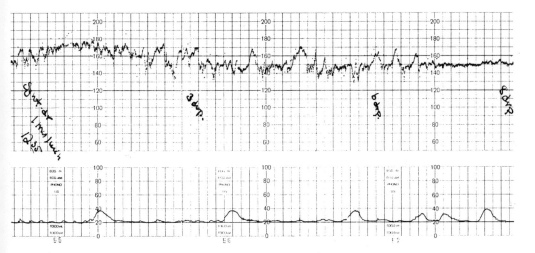

Fig. 14.27 A negative stress test (a reactive FHR and no decelerations) with an oxytocin infusion of 8 mU/min.

The OCT is contraindicated in women at risk of preterm labour or scar rupture. The indications for an OCT are not well defined and have varied in recent years. Some use the test routinely in high-risk patients but most obstetricians now only consider testing pregnancies with FHR abnormalities on the non-stress test, particularly if post-term or with fetal growth retardation. Testing placental function in pregnancies with occasional late decelerations in an otherwise reactive CTG, or before cervical ripening with prostaglandin in cases with placental insufficiency, are possible indications, although some would argue that the process of induction is itself a contraction stress test. There is no reason to use the OCT in normal pregnancies or after a reactive NST. The OCT has never been popular in the UK.

One of the problems with the OCT is that the amplitude of induced contractions cannot be quantified. It is, therefore, not certain that repeated tests expose the fetus to the same stress. Such uncertainty is illustrated by the fact that in some cases with a positive result the FHR reverts to normal after amniotomy. Similarly, after a positive test, there are many reports of normal labour and vaginal delivery. False negative tests have also been reported. Early experience with OCT testing was to repeat the test weekly if the result was normal. However, as with non-stress testing, such a rigid policy may not be appropriate. We found that 9 of 70 women (13 per cent) had late decelerations on the NST or in labour within 7 days of a negative OCT.

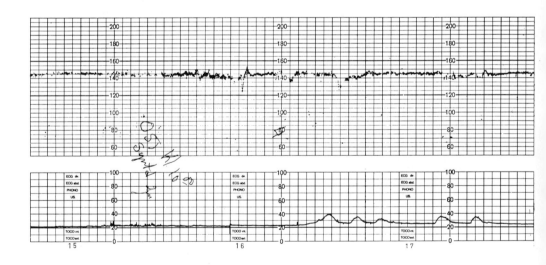

A

Fig. 14.28 (caption opposite).

The oxytocin challenge test (OCT)

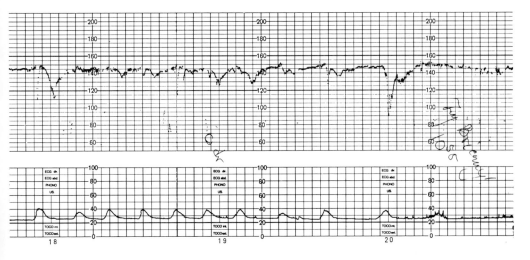

B

Fig. 14.28 An OCT at 32 weeks because of suspected intrauterine growth retardation. A, A non-reactive CTG soon after the onset of the oxytocin infusion (between marks 15 and 16). A uterine response can be seen within 10 min; B, small, repetitive, late decelerations indicate a positive result. The oxytocin infusion was stopped at mark 19 but the contractions continued. Terbutaline was given (0.25 mg i.v.). An emergency Caesarean section was performed after the test and Apgar scores were 4, 6, and 8 at 1, 5, and 10 min. The birthweight was 1000 g.

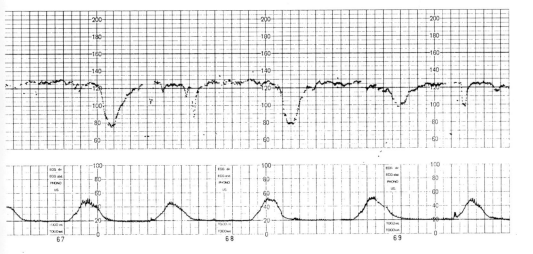

Fig. 14.29 Severe intrauterine growth retardation at 34 weeks. A positive OCT at 4 mU/min of oxytocin showing repetitive late decelerations and low variability.

Summary

Antepartum FHR monitoring should be performed for diagnostic purposes in the light of clinical circumstances. Most tests are reactive, even in high-risk pregnancies, but this only implies fetal adaptation at that point in time. Non-reactive tests are often due to quiet fetal behaviour and the record should be prolonged in such circumstances.

Compared to non-stress testing, the OCT is laborious and costly. The average time for the procedure is more than 2 h. Its revelance to clinical practice remains unclear and it is rarely used in the UK.

15
Terminal fetal heart rate patterns

Introduction

A major objective of fetal monitoring is to prevent intrauterine death and to detect fetal distress. Antepartum stillbirths (fetal deaths after 24 weeks) still contribute significantly to perinatal mortality, whereas rates of intrapartum stillbirth at term have fallen significantly since the introduction of continuous FHR monitoring into clinical practice.

Incidence of stillbirth

Estimates of the rate of hypoxic fetal death during labour were one in 700 during the 1970s, when intermittent auscultation was the normal method for monitoring the FHR. The FHR may be normal prior to fetal death but loss of short-term variability, the absence of accelerations and small, shallow, late decelerations will not be identified by intermittent auscultation (Fig. 15.1). The baseline FHR may remain normal until a terminal bradycardia occurs.

There were no intrapartum fetal deaths in a low-risk population of 5037 women who had routine continuous FHR monitoring during labour in Lund between 1977 and 1978 (Westgren et al. 1980). Others have also reported significant reductions in rates of fetal death due to hypoxia during labour as a result of continuous FHR monitoring. However, whether the reduction in fetal mortality can be attributed entirely to the method of monitoring, or whether other changes relating to the quality of intrapartum care are associated with such practice, remains a matter of debate. In Dublin, a large randomized controlled trial of continuous FHR monitoring in labour did not show a lower rate of intrapartum stillbirths in the monitored group compared with the existing very low rate of 0.5 per 1000 in the control population studied (MacDonald et al. 1985). Continuous FHR monitoring in labour also has its drawbacks (see Chapter 18).

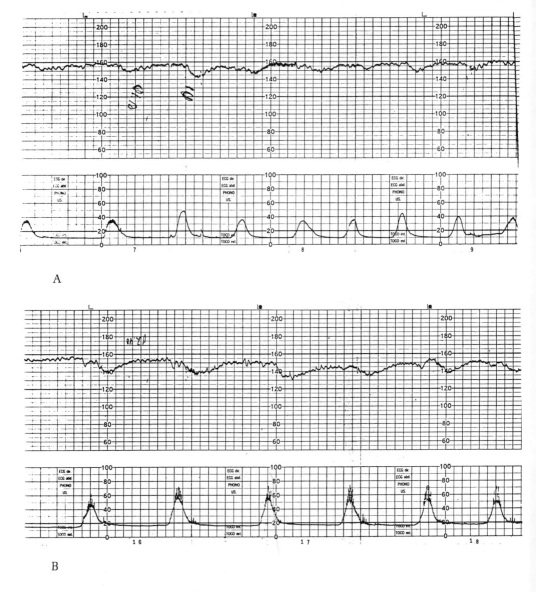

Fig. 15.1 (caption opposite).

Aetiology of stillbirths

Fetal deaths still occur as a result of anomalies, infection, and accidents such as placental abruption, cord prolapse, and uterine rupture. Sepsis at,

Aetiology of stillbirths

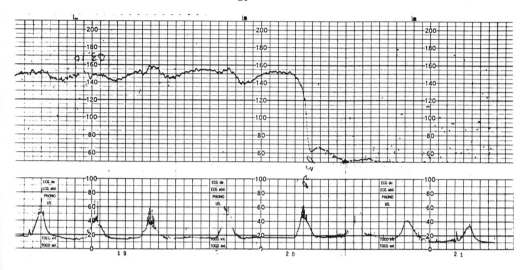

C

Fig. 15.1 A primiparous woman in spontaneous labour at term after a normal pregnancy. The amniotic fluid was lightly meconium-stained. The fetus died 3 h after admission to the labour ward. External record of uterine activity; FHR by scalp electrode. A, The appearance soon after admission to the labour ward was of a mild tachycardia (155–160 bpm) with low variability and absent short-term variability. Mild late decelerations can be seen; B, 70 min later, a similar FHR pattern; C, this is a continuation of B, which ends in a terminal bradycardia for few minutes before the fetus dies. At autopsy, signs of severe fetal hypoxia were evident. This is an example of misinterpretation of the FHR pattern where the presence of some long-term variability was interpreted as normal. The meconium-stained liquor was an additional warning signal.

or soon after, birth may result in a sudden and rapid demise of the neonate (Fig. 15.2). Unexplained intrauterine fetal death in labour, shortly after a reactive FHR pattern, has also been reported.

Intrauterine death in the antenatal period remains a problem and fetal hypoxia is implicated in many cases. A reactive CTG is not a guarantee against subsequent fetal death, even within a few hours or days (Fig. 15.3), and so it is only appropriate to consider the test as a diagnostic one for immediate assessment of fetal well-being. In 13 reports of fetal deaths before labour the FHR pattern was only normal in two of 86 patients. In another series, 49 perinatal deaths occurred after a reactive CTG of which acute events, not related to hypoxia, were responsible in 26 (53 per cent).

Cardiotocography is the most common method for monitoring fetal well-being in the antenatal period. However, the method should be reserved for

258 Terminal fetal heart rate patterns

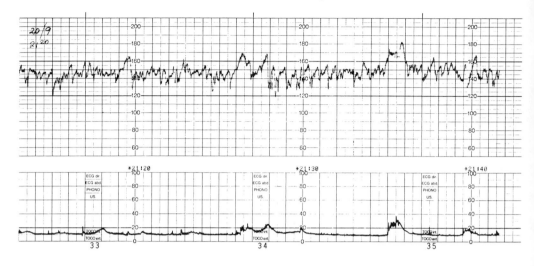

A

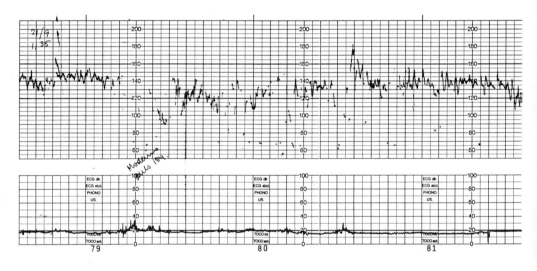

B

Fig. 15.2 Ruptured membranes for a few days at 32 weeks. A, A normal FHR pattern 18 h prior to fetal death; B, 3 h before fetal death the CTG was still reactive, although the quality of the recording is not optimal due to bouts of maternal shivering. Autopsy showed massive bacterial growth (*Staphylococcus aureus*) in the fetus.

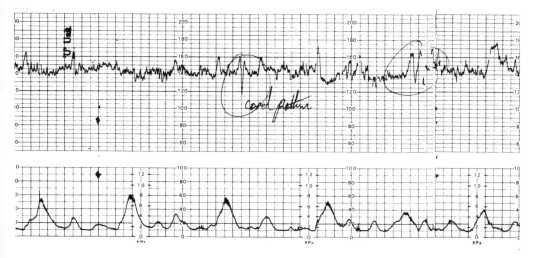

Fig. 15.3 A primigravida at term with hypertension. An ultrasonic scan just prior to the CTG had shown a normal fetus and a normal amount of amniotic fluid. The CTG is reactive. A type 0 dip (a small variable deceleration of very short duration) is marked in the middle of the tracing. This type of dip may be seen with cord compression or fetal hiccups. Fetal death occurred 16 h later.

women with pregnancy complications. In Lund during 1980, 812 women with pregnancy complications were monitored daily with a CTG. Practically all cases with ominous FHR patterns were already high risk. Only 1.1 per cent of low-risk women had a pathological record and there were no antenatal deaths in this group. Two intrauterine deaths occurred, both in high-risk pregnancies. In one, the FHR pattern (absent baseline variability and repetitive late decelerations) was misinterpreted; in the other, the fetus was considered too small at 26 weeks gestation and intervention was withheld (Fig. 15.4). Thus, 'CTG clinics' for outpatients are not indicated, even for weekly testing. Such a policy will divert resources from more appropriate, complicated, cases without benefit for mother and fetus.

Terminal FHR patterns

Although there is no specific terminal pattern, reports suggest that certain FHR changes occur more often than others. However, it is emphasized that no reliable prediction can be made as to how long the fetus will survive after a terminal FHR pattern is found; it may be less than a few hours or it may be as long as several days.

260   Terminal fetal heart rate patterns

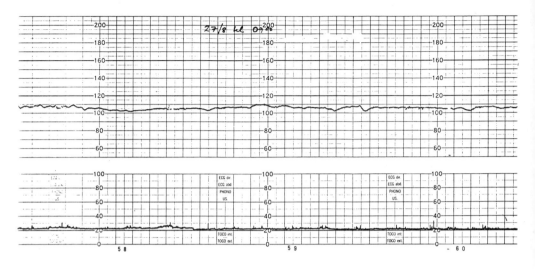

A

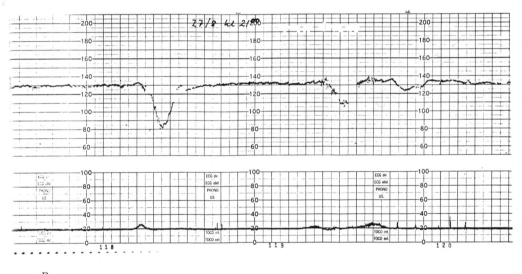

B

Fig. 15.4 (caption opposite).

Terminal bradycardia is a late change and is often preceded by tachycardia and loss of variability. Absence of variability, with or without tachycardia, is a common feature in all cases of intrauterine fetal death and both short-term and long-term variability is absent. A normal FHR with loss of variability may

Terminal FHR patterns

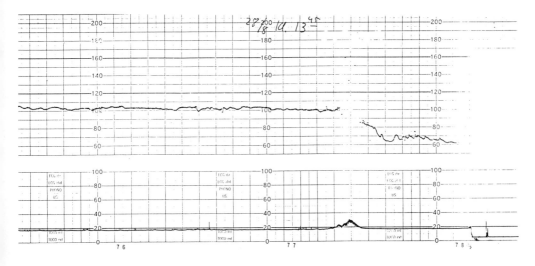

C

Fig. 15.4 A primipara at 26 weeks with a severely growth retarded fetus. A, A CTG with moderate bradycardia and absent variability; B, 12 h later, late decelerations are seen with Braxton Bicks tightenings; C, 16 h later, a progressive bradycardia with absence of baseline variability occurred before fetal death.

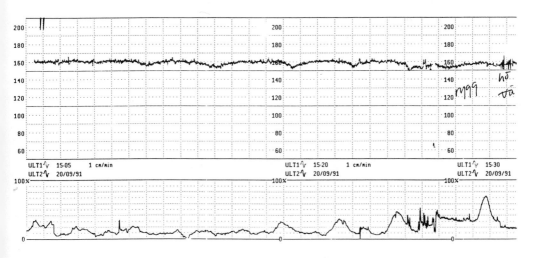

Fig. 15.5 A primipara at 41 weeks after an uneventful pregnancy. She reported a cessation of fetal movements lasting 2 days and was admitted for a CTG. The tracing shows a tachycardia (160 bpm), very low baseline variability and occasional late decelerations. Intrauterine fetal death occurred 2 h later. Autopsy showed signs of severe intrauterine asphyxia.

also precede fetal death (Fig. 15.5). In preterm fetuses, pronounced variable decelerations with overshoot and loss of baseline variability may also be seen (Fig. 15.6).

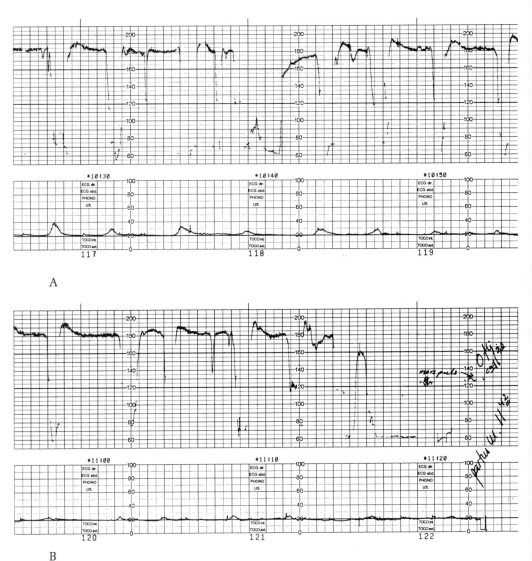

Fig. 15.6 A multipara at 26 weeks with prelabour rupture of the membranes for 7 days and severe chorioamnionitis. A, Recording of the FHR 60 min prior to fetal death showing pronounced variable decelerations with overshoot and absent long-term variability; B, the pattern was followed by a terminal bradycardia and fetal death.

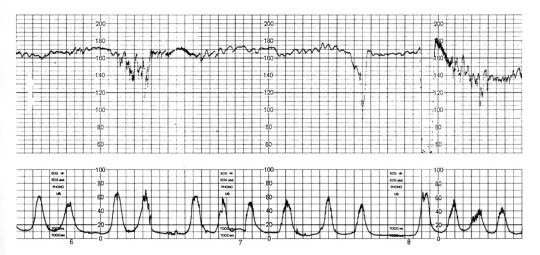

Fig. 15.7 A tachycardia (FHR 170 bpm) with absent short-term variability and sporadic, unclassified, decelerations.

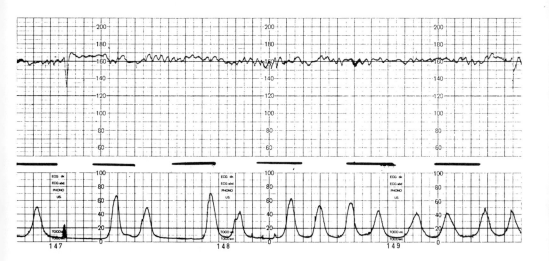

Fig. 15.8 Tachycardia and sinusoidal-like FHR variability preceding fetal death.

Pronounced late (or variable) decelerations are common with contractions. A mixture of late and variable decelerations, or pronounced decelerations with an atypical waveform can also be seen. These decelerations start at, or after, the peak of the contraction but may have a 'variable' waveform (Fig. 15.7). Late or variable decelerations may then disappear, leaving an absence of variability as the only abnormal feature of the record. This may represent a total lack of fetal reflex control due to hypoxic damage to the fetal brain stem and, possibly, to the heart. A sinusoidal FHR pattern may also occur (Fig. 15.8). In the antenatal period without contractions, absence of baseline variability may be the only abnormality.

Summary

Terminal FHR abnormalities are usually seen antenatally in pregnancies complicated by chronic placental insufficiency. If unrecognized, such pregnancies may enter labour, at which time even the reduction in uteroplacental perfusion associated with normal uterine activity may be sufficient to reduce fetal oxygenation below a level compatible with survival. The precise nature of the reduction in fetal oxygen supply, and its time course, are important variables in determining the precise FHR response.

16
Effects of drugs on the fetal heart rate pattern

Local anaesthetics

The FHR can be affected by local anaesthetics both indirectly (by affecting the maternal circulation) and directly (by a depressive action on the fetal heart). Paracervical and epidural block, used for pain relief during labour, frequently induce FHR changes.

Paracervical block

The frequency of FHR changes, most often bradycardia or prolonged decelerations, varies according to the type and concentration of the local anaesthetic. Typically, an abrupt bradycardia is seen about 10 min after the block is given, although late decelerations may occur. The FHR may fall to between 60 and 70 bpm with reduced variability (Fig. 16.1). The FHR

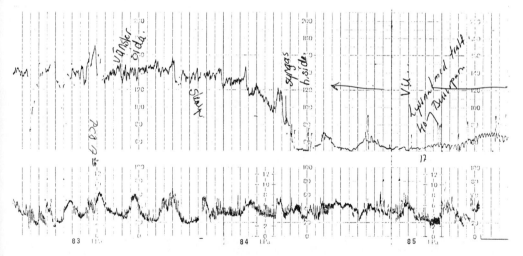

Fig. 16.1 Fetal bradycardia (50–60 bpm) with loss of variability 10 min after insertion of a paracervical nerve block (at mark 83). An emergency Caesarean section was performed.

generally recovers within about 10 min and may be followed by a transient, compensatory, tachycardia. Fetal pH usually remains normal if the bradycardia recovers within 10 min.

The origin of the FHR response is not clear but a combination of factors is likely. Direct depression of the fetal heart is possibly an important factor, which follows rapid absorption of the drug into the fetal circulation. Other factors, such as an increase in uterine activity and constriction of the uterine artery causing reduced placental circulation, may also be relevant. Local anaesthetics are known to cause vascular constriction and an effect on the uterine circulation is likely without a significant change in maternal systemic blood pressure. Thus, the FHR effect is likely to be a response to transient hypoxia resulting from reduced maternal placental perfusion.

Epidural block

Severe bradycardia occurs less frequently after epidural block than after paracervical block, particularly if a fall in maternal blood pressure is avoided. However, late, variable, or combined decelerations may appear without a measurable reduction in maternal arm blood pressure (Figs 16.2 and 16.3). Epidural analgesia results in a sympathetic block with vasodilatation in the lower body, diminishing venous return. Perfusion pressure in the uterine vasculature falls if maternal blood pressure falls. After epidural block the femoral artery blood pressure may be 20 mmHg lower than pressure in the brachial artery, suggesting the possibility of reduced placental perfusion. A greater predisposition to aortocaval compression in the supine position may also be a contributory factor as is abnormal uterine activity concomitant with the analgesia.

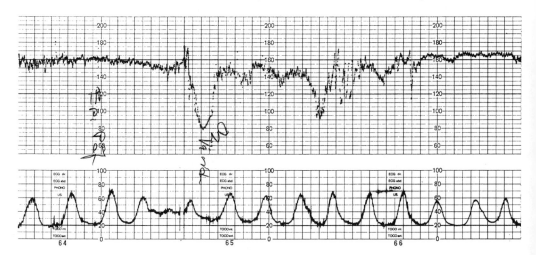

Fig. 16.2 Late decelerations 5 min after an epidural block.

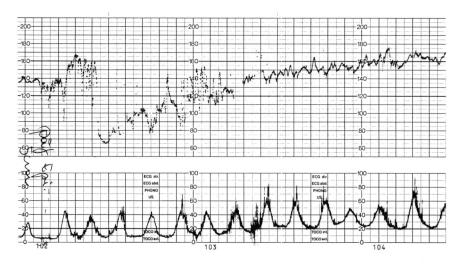

Fig. 16.3 A prolonged deceleration after epidural block.

Up to 10 per cent of labours receiving epidural analgesia will demonstrate FHR decelerations within 30 min. Oxytocin infusion is an added predisposition and should be administered with caution. An epidural, particularly in an induced labour, is a well accepted indication for continuous FHR monitoring during labour. Maternal temperature may rise after administration of an epidural block resulting in a fetal tachycardia (see Chapter 6).

Oxytocin

Oxytocin may directly reduce long-term variability and, to some extent, short-term variability. However, the main effect on FHR is indirect via stimulation of uterine activity and reduction of uteroplacental perfusion (see Chapter 4).

β-receptor agonists

A β-mimetic drug given intravenously to the mother usually causes a maternal tachycardia greater than 100 bpm. Drugs such as terbutaline and salbutamol cross the placental barrier and often induce a fetal tachycardia when given by infusion for the inhibition of preterm uterine activity. However, other reasons for a fetal tachycardia, such as fetal infection, need to be excluded.

Maternal hypotension during infusion of a β-mimetic drug may also result in FHR changes, although this seldom occurs if such drugs are given with adequate caution and supervision.

β-receptor blockers

Drugs of this type may be used for antihypertensive treatment in pregnancy, sometimes in combination with hydralazine (see Fig. 11.5). Propranolol, a non-selective β-blocker, has been reported to cause fetal bradycardia (Fig. 16.4) and has been associated with placental insufficiency.

Modern β-blockers are of two types: cardioselective (such as atenolol and metoprolol), which mainly block β-receptors, and non-cardioselective with intrinsic sympathomimetic activity (such as pindolol and labetalol), which also have some α-blocking activity. The influence of these two types of drug on FHR

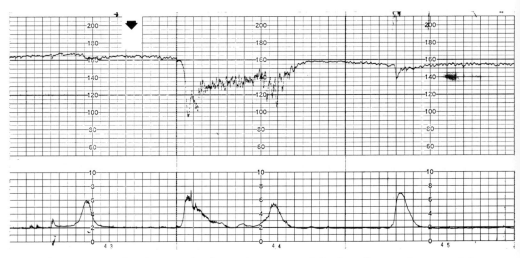

Fig. 16.4 Decelerations 3 min after an intravenous injection of 2 mg of propranolol (at the arrow).

Table 16.1 Ominous fetal heart rate changes in 200 matched pairs

	Hypertensives (n = 200)		Controls (n = 200)	
	No.	%	No.	%
Complicated baseline*	17	8.5	4	2.0
Ominous decelerations				
late	9	4.5	4	2.0
combined	9	4.5	3	1.5
pronounced variable†	6	3.0	2	1.0
Total	41	20.5	13	6.5

*Tachycardia or bradycardia with silent pattern of decelerations (including prolonged decelerations).
†Lost beats >60 bpm, duration >60 s (Montan and Ingemarsson 1989).

is different. Cardioselective drugs, such as atenolol, have been reported to reduce the amplitude and number of FHR accelerations and reduce the baseline rate and long-term variability. No such changes occurred in hypertensive women treated in the third trimester with pindolol. Possible effects on the FHR pattern during labour after treatment with β-blockers in pregnancy is not clear.

Women with hypertension had ominous FHR changes in labour more frequently than a group of control women matched for age, parity, and gestational age (Table 16.1). Ominous FHR records were seen in 29 per cent of women treated with a β-adrenergic receptor blocker (atenolol) and in 16.6 per cent of patients without such treatment. This is in line with reported adverse effects of these drugs in animal experimental studies. However, these results should be interpreted with caution because treatment was given to the more serious cases.

Atropine

Atropine is an anticholinergic drug with a strong vagolytic effect. Early decelerations due to vagal reflex activity will be abolished and late decelerations only modified (see Chapter 8). This should be remembered if decelerations 'improve' after administration of atropine as premedication.

Narcotic sedatives, hypnotics, and analgesics

These drugs cross from the mother to the fetus and exert similar effects. The fetal response will depend on the type of drug and method of administration. Pethidine is commonly used for pain relief in labour and its effects are well known. One dose of 50–100 mg of pethidine i.m. will affect the FHR pattern in three ways. First, there will be a reduction in FHR variability due to fetal quiescence (Fig. 16.5). The maximal effect will appear 40 min after the injection and long-term variability is affected more than short-term variability. The variability usually returns after about 60 min but not invariably so (Fig. 16.6). Second, the baseline FHR often decreases by 10–12 bpm. This decrease may be independent of the influence on variability (Fig. 16.7). Third, the number and amplitude of accelerations are often reduced (Fig. 16.7).

The FHR pattern changes from a reactive to a non-reactive one and the alterations resemble a change in behavioural state from activity to quiescence. Such an effect on fetal behaviour is to be expected and may explain the variations in effect on these parameters of the FHR. Sometimes all three parameters are reduced, sometimes only one or two.

Effects of drugs on the fetal heart rate pattern

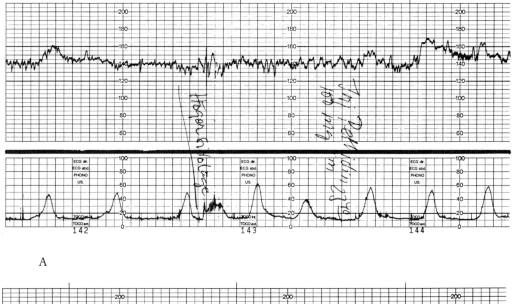

A

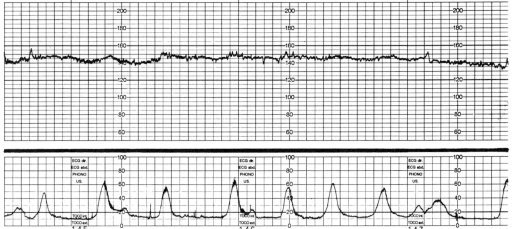

B

Fig. 16.5 (caption opposite).

Interpretation of the FHR may be more difficult after administration of narcotic drugs. It is therefore sensible to record the fetal heart rate before giving these drugs to avoid uncertainty about the origin of FHR influence. It is possible to given an antidote if necessary in difficult situations.

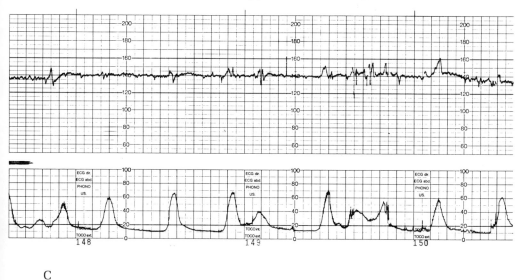

C

Fig. 16.5 A primipara at term in spontaneous labour. A, Baseline FHR 140 bpm with normal variability and a reactive pattern. An injection of 100 mg pethidine i.v. was given (between marks 143–144); B, after the pethidine, reduced variability and reduced reactivity occur without change in the baseline FHR; C, 40 min after the injection, reduced variability persists.

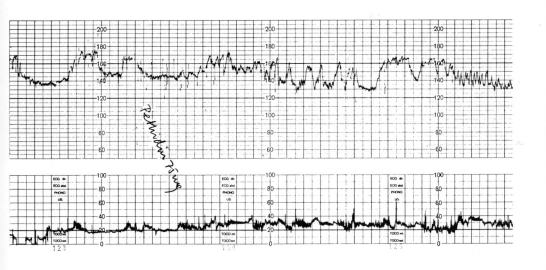

A

Fig. 16.6 (caption overleaf).

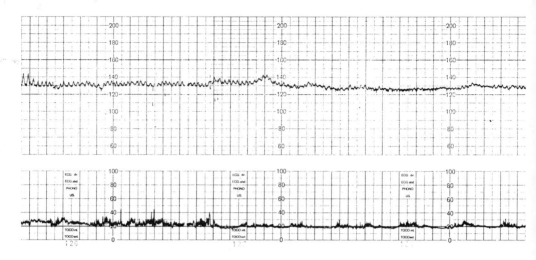

B

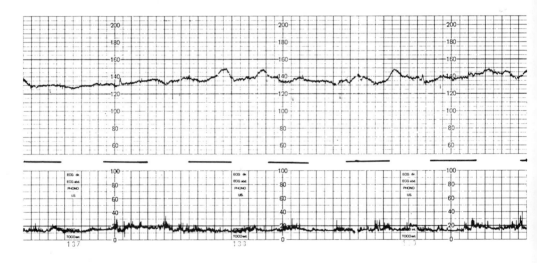

C

Fig. 16.6 A primipara at term in spontaneous labour. A, Pethidine 75 mg i.m. was given at the beginning of the recording. The baseline FHR is 140 bpm with normal variability and reactivity; B, 20 min after the injection, the FHR is 130 bpm with reduced variability and absence of accelerations; C, 2 h and 20 min after the injection, variability is almost absent. Delivery occurred 2 h and 10 min later. Apgar scores were normal.

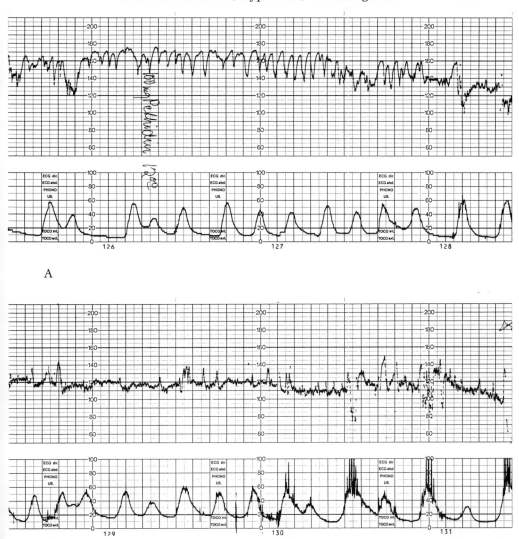

Fig. 16.7 A multipara at term in spontaneous labour. A, 10 mg pethidine i.v. was given at the beginning of the record. The baseline FHR is 160 bpm with frequent accelerations; B, 15 min after the injection the FHR fell to just below 120 bpm; the amplitude of accelerations and variability are both reduced.

17
Admission Test and fetal stimulation

Introduction

A major aim of continuous electronic fetal monitoring in labour is to detect signs of possible fetal distress as early as possible. Commencement of such monitoring soon after admission in labour is designed to screen for such signs immediately using the principle that labour acts as a spontaneous stress test.

Principles of the Admission Test

Babies may be born asphyxiated after a short recording with an ominous FHR pattern (Fig. 17.1). In some cases (such as this one) the mother may have arrived at the hospital several hours earlier but a number of preparations for labour had to be completed before monitoring was started. Given the fact that uterine contractions reduce uteroplacental circulation and reduce fetal oxygen delivery, a short (up to 30 min) record of the FHR immediately after admission to the labour ward—an Admission Test—was advocated. By this means, intrauterine hypoxia, already present on admission, may be detected, thereby allowing appropriate measures to confirm hypoxia (scalp blood sampling) or act on it (delivery) to be undertaken. It was also felt that, if a normal CTG on admission has some predictive value for fetal well-being for the next few hours of labour, continuous FHR monitoring might not be necessary thereby allowing the mother to remain ambulant with intermittent auscultation.

As discussed in Chapter 14, chronic placental insufficiency with a small margin of fetal reserve means that the fetus has probably adapted satisfactorily to the reduced uteroplacental perfusion. However, once uterine contractions start with the onset of labour then the effects of acute on chronic hypoxia may be seen on the FHR record. The Admission Test was designed to serve as a screening test for such cases.

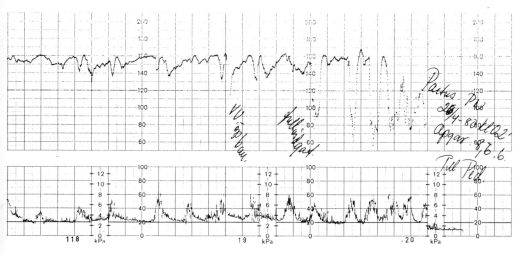

Fig. 17.1 A primipara, 34 years of age, in spontaneous labour after an uneventful pregnancy of 41 weeks. On admission with intact membranes, the cervix was 4 cm dilated and the fetal head engaged. The first record of the FHR was done 1.5 h after admission and lasted 25 min, as shown. The baseline FHR was 160 bpm with absent short-term variability and decelerations. Thick meconium (and blood) appeared at the time of amniotomy. A rapid vaginal delivery followed and the third stage was accompanied by signs of placental abruption. Apgar scores were 8, 6, and 6 at 1, 5, and 10 min. The baby died the following day as a result of the severe asphyxia.

Interpretation of the Admission Test

Normal/reactive

The presence of two accelerations in 20 min is a reactive test result (Fig. 17.2). However, records with no accelerations but normal baseline rate and normal baseline variability are not abnormal. Also acceptable is a normal baseline rate with accelerations despite early decelerations.

Equivocal/suspicious

A non-reassuring result is a normal baseline rate with no accelerations and low baseline variability, an abnormal baseline rate (above 160 bpm) with no accelerations, or variable decelerations without ominous signs (Fig. 17.3).

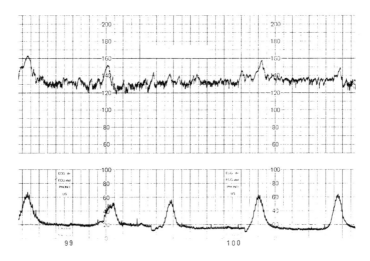

Fig. 17.2 A reactive Admission Test.

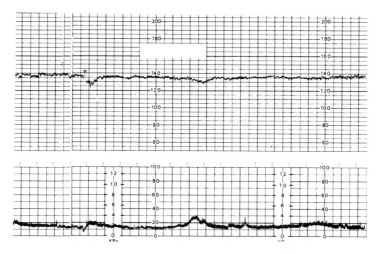

Fig. 17.3 An equivocal Admission Test.

Abnormal/ominous

A worrying result would be a baseline variability less than 5 bpm and an abnormal baseline rate, especially with repeated late decelerations. Repeated variable decelerations with any ominous features (duration greater than 60 s and decelerating more than 60 beats from the baseline FHR, rebound tachycardia, overshoot, slow recovery, reduced variability within the

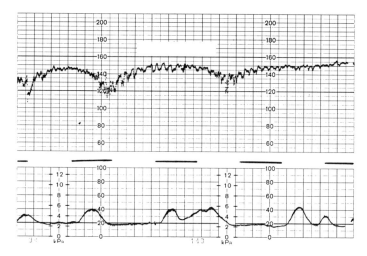

Fig. 17.4 An ominous Admission Test.

decelerations, or late component) would also be a cause for concern. A bradycardia below 100 bpm or prolonged decelerations are also abnormal (Fig. 17.4).

Use of the Admission Test

The use of the Admission Test for screening has been studied in 1041 women admitted in the first stage of labour after 34 weeks of pregnancy. Women with spurious (false) labour who delivered more than 24 h after admission were excluded. On admission to the labour ward a 20 min CTG was obtained. The results of this test were concealed from the clinicians managing labour: only the lamp indicating optimal recording of the FHR was visible. The trace was then placed in an envelope until the outcome of labour was known so as not to influence management. Fetal distress was considered present when subsequent ominous FHR changes resulted in a Caesarean section or forceps delivery, or if the baby was depressed (Apgar score less than 7 at 5 min) after spontaneous delivery.

Reactive Admission Tests were found in 94.3 per cent of the 20 min records (Table 17.1) and fetal distress in labour was uncommon in this group (1.3 per cent). Equivocal tests and ominous tests were associated with much higher rates of fetal distress. Thick meconium was more commonly associated with an abnormal FHR on admission. Of a total of 14 women with thick meconium at admission or at amniotomy, ten had a reactive test and only one Caesarean section was performed in this group. Four patients had a non-reactive test (equivocal 2, ominous 2) and all had an abdominal delivery for fetal distress.

Table 17.1 The Admission Test in relation to fetal distress (Ingemarsson et al. 1986)

	Outcome of Admission Test		Fetal distress	
	No.	%	No.	%
Reactive/normal	982	94.3	13	1.3
Equivocal	49	4.7	5	10.0
Ominous	10	1.0	4	40.0

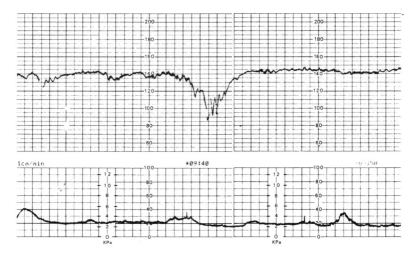

Fig. 17.5 An ominous Admission Test 3 h prior to intrapartum fetal death. The baseline FHR is 145 bpm with absent variability and late decelerations.

The only intrapartum fetal death occurred among the four patients with an ominous Admission Test (Fig. 17.5). Labour was spontaneous and normal heart tones were intermittently auscultated. At amniotomy, thick meconium was detected and the FHR could not be heard subsequently. An ultrasound scan confirmed intrauterine death. At autopsy there were signs of severe fetal asphyxia in a normally formed infant. Had the ominous Admission Test result been known then appropriate intervention is likely to have produced a different result.

In total, 22 patients had fetal distress and the majority of these (13) developed this complication despite a reactive Admission Test. Of interest is the length of time from admission to intervention (or delivery) in this group. Only three of the 13 patients were delivered within 6 h of admission; one had a Caesarean section for cord prolapse 1.5 h after admission, and the other two were delivered 3 and 4 h after admission, respectively. The Admission Test cannot be expected to predict fetal distress that develops several hours later in labour if fetal condition was satisfactory on admission. Thus, rather like pregnancy, a negative test result has limited prognostic value but a positive result is justification for further action. However, because of the low prevalence of abnormal tests and fetal distress, statistical analysis of reactive and ominous test results indicates a high predictive value for fetal well-being of a normal test (98.7 per cent) and a high specificity (99.4 per cent), but a rather low predictive value of an abnormal test (40.0 per cent) and a low sensitivity (23.5 per cent).

In summary, the results suggest that the Admission Test can detect fetal distress already present on admission. Unnecessary delay in further action can, therefore, be avoided in such cases (see Figs 5.14, 17.6 and 17.7). The test has some predictive value for fetal well-being for the subsequent few hours of labour but changes in clinical circumstances may interfere with this. The need for continuous monitoring for a while after a normal Admission Test is low if no further risk factors are observed. Results of the Admission Test are summarized in Fig. 17.8.

The subsequent duration of labour is an important risk factor for the development of fetal distress following a reactive CTG on admission in labour. Krebs *et al.* (1979) found the frequency of low Apgar scores to be high (69.6 per cent) if the first 30 min of the labour CTG was abnormal compared with records that were normal (2.7 per cent) or suspicious (15.8 per cent). Unfortunately, the time relation between admission and initial FHR monitoring is not stated in their study. In the Dublin FHR monitoring study (MacDonald *et al.* 1985), which

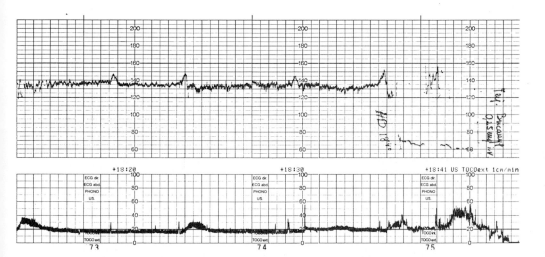

Fig. 17.6 A reactive Admission Test terminates with a severe bradycardia at the time of spontaneous rupture of the membranes. A cord prolapse was diagnosed at vaginal examination. Terbutaline 0.25 mg i.v. was given and an emergency Caesarean section was performed. The fetus was in good condition at birth with normal Apgar scores.

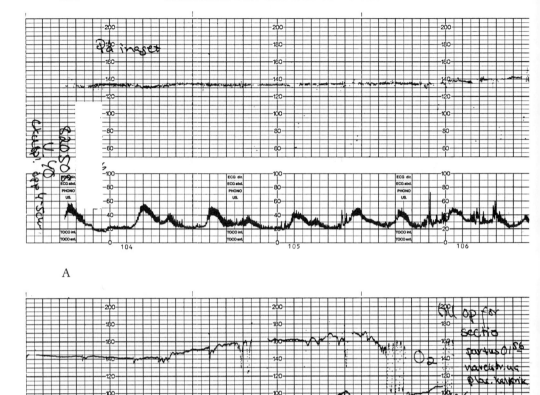

A

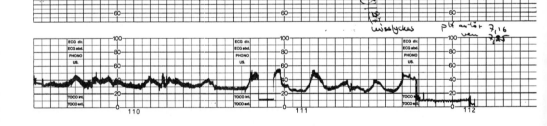

B

Fig. 17.7 A primipara in spontaneous labour at term (cervix effaced, and dilated between 4 and 5 cm). A, The Admission Test shows absence of baseline variability and is non-reactive. The woman was taken immediately to a labour room for amniotomy and application of a fetal scalp electrode for internal monitoring; B, The FHR after amniotomy had absent variability. At fetal scalp blood sampling a prolonged deceleration occurred. An emergency Caesarean section was performed and the Apgar scores were normal. The umbilical cord artery pH was 7.16; venous pH was 7.25. This is an example of mild fetal distress in a low-risk pregnancy detected immediately after admission.

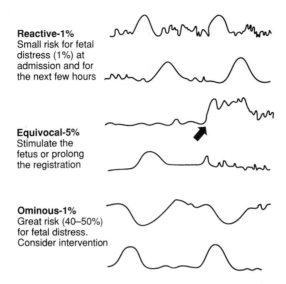

Fig. 17.8 A summary of fetal outcome and Admission Test patterns.

compared continuous FHR monitoring with intermittent auscultation, the rate of neonatal seizures was significantly higher in the auscultation group (8.5 per cent) compared with the electronically monitored group (2.4 per cent) and further analysis showed that these patients often had experienced labour lasting more than 5 h and had required more oxytocin.

Fetal stimulation in labour

Evidence of fetal reactivity (variability and accelerations) in response to fetal movements is the basis of the assessment of fetal well-being. The likelihood of this pattern being present with fetal hypoxia is very low. In the antenatal period, some 10 per cent of records are interpreted as non-reactive and at admission in labour about 5–6 per cent of tests will be regarded as equivocal or ominous. During labour about half of all records show FHR changes, although the vast majority are innocuous. Determination of fetal pH is an important means by which fetal condition can be further evaluated in cases with abnormal FHR changes.

Sometimes it is difficult, or even impossible, to collect blood from the fetal head during labour, particularly when the head is high and mobile and the cervix is not dilated. In such circumstances, FHR accelerations are a reassuring sign. If spontaneous accelerations are not present, the possibility of provoking fetal activity with its FHR reactivity has been recently evaluated. The fetal response to various stimuli, such as light, sound, vibration, and manipulation, is usually with an FHR acceleration. Such reactions suggest a well-oxygenated fetus.

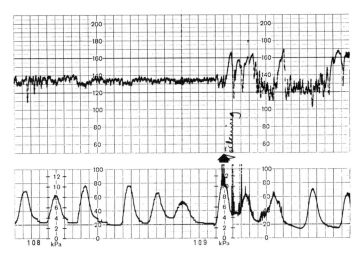

Fig. 17.9 An external CTG with a baseline FHR of 135 bpm and reduced variability. A positive fetal response to manipulation (at the arrow) indicates fetal arousal and the CTG subsequently indicates the fetal active state with accelerations.

Manipulation

Manual manipulation of the uterus has been used for many years as a means of arousing the fetus and converting a fetal quiet state to fetal activity. Accompanying such a state change will be the FHR differences between a non-reactive and a reactive CTG (Fig. 17.9). Fetal arousal is considered a reassuring sign, indicating health; failure to respond is considered to be a risk factor for the presence of fetal hypoxia.

The relationship between manipulation and fetal response is not straightforward. Manipulation affects fetal behavioural states to a smaller extent than expected and randomized studies have suggested that the change in fetal activity following manual stimulation occurs by chance (see Fig. 5.24).

The scalp stimulation test

The FHR often accelerates when the scalp is incised during fetal blood sampling (see Figs 5.31C and 17.10). Studies have shown that such accelerations are highly predictive of the absence of fetal acidosis. However, the absence of an FHR acceleration following scalp stimulation does not indicate acidaemia, as most will also be normal. Nevertheless, acidaemic fetuses will be in this group of non-responders. Scalp stimulation has been performed with Allis tissue forceps and digital pressure at vaginal examination. It is possible that simple, non-invasive tests like these are of value. Positive fetal responses are reassuring and intervention may be avoided if scalp blood is difficult or impossible to obtain.

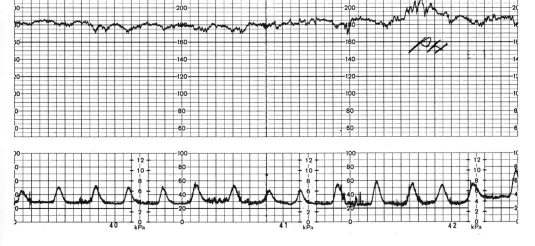

Fig. 17.10 A primipara at 40 weeks gestation. The non-reactive tachycardia was an indication for a fetal blood pH sample at which time a FHR acceleration occurred. The pH value was normal (7.39).

Vibroacoustic stimulation

Vibroacoustic stimulation of the fetus through the maternal abdomen is another method by which fetal state may be changed. Fetal arousal is accompanied by a transient tachycardia and persistence of the active state with FHR variability and accelerations. The vibration component of the stimulation is probably more important than the sound.

Background noise within the uterus has been assessed by placing intrauterine microphones within the uterine cavity. Sound levels of 85–95 dB have been reported, with peaks at about 300 ms after the R wave of the maternal ECG. Thus, much of this noise comes from maternal vascular turbulence, as well as from muscle activity. External noise is attenuated by the abdominal and uterine walls, particularly for frequencies of sound above 1000 Hz, when attenuation can reach 50 dB (2000 Hz at 120 dB level). Therefore, the fetus may not perceive noise of high frequency.

An artificial (electronic) larynx is used to arouse the fetus and induce a reactive CTG. This may be indicated as a result of a non-reactive CTG, an equivocal Admission Test or a suspicious intrapartum CTG, particularly if the only abnormality is prolonged reduction of variability for more than 60 min (see Figs 6.13, 7.10 and 17.11). Its use during pregnancy shortens antepartum testing time, although this has not been shown to have greater value than waiting for spontaneous fetal activity. It is a rapid procedure, requiring only 10 min testing

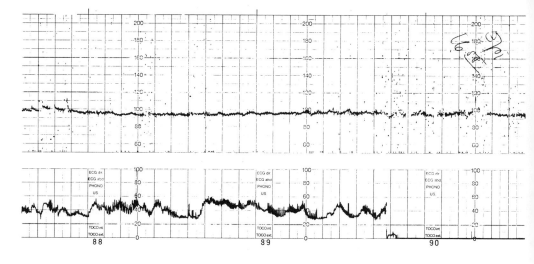

A

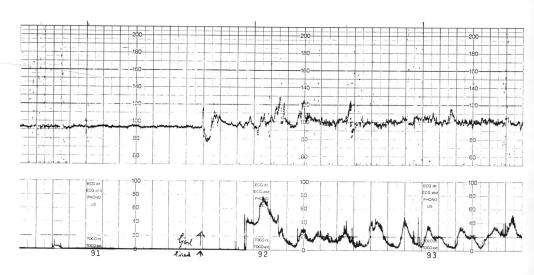

B

Fig. 17.11 A primipara at 42 weeks in spontaneous labour. A, A baseline bradycardia of several hours duration with absence of variability; B, acoustic stimulation (at the arrow) resulted in a reactive pattern with variability and accelerations. Such a response to stimulation occurred on two further occasions during the next 3 h until delivery. The fetus was in good condition at birth with normal Apgar scores.

time if the fetus responds. The positive predictive value for fetal compromise appears to equal that of the normal CTG. The fetal movements that result from such stimulation are usually perceived by the mother.

Acoustic stimulation may also be used in conjunction with an equivocal Admission Test, which is found in about 5 per cent of all cases. A non-reactive FHR pattern, which shows a deceleration in response to stimulation, may have a poor outcome. Similar findings were found when acoustic stimulation was used together with the Admission Test. The risk for fetal distress in labour was high (three out of four) after an ominous Admission Test and a decelerative FHR response to acoustic stimulation (see Figs 5.14 and 17.12).

In labour, a positive fetal response to acoustic stimulation is mostly associated with normal fetal scalp blood pH in cases with suspicious or ominous traces. However, there is some evidence that a fetal response is possible with mild acidaemia, providing it is respiratory rather than metabolic. Perhaps a fetus with a respiratory acidosis is not as hypoxic as one with a metabolic acidosis and is therefore likely to exhibit a response to external stimulation.

Concern about possible harm to the fetus as a result of vibroacoustic stimulation *in utero* has been expressed. However, long-term follow-up has not shown any evidence of damage to the neurodevelopmental integrity or hearing ability of children exposed to the test before birth.

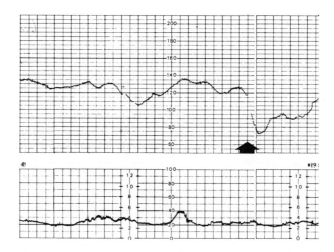

Fig. 17.12 A terminal FHR pattern on admission. Acoustic stimulation (at the arrow) was followed by a bradycardia. Intrauterine death occurred soon after this record.

18
Clinical considerations

Introduction

Continuous FHR monitoring was introduced more than two decades ago for selected 'high-risk' cases; antepartum CTG records were uncommon. Within 15 years of its introduction, routine electronic monitoring of all women in labour was common in many Western centres. During the 1980s, monitors were also being obtained by countries in the developing world, despite limited financial resources. At this time, experience was resulting in a growing realization in the West that the contribution by such technology towards a reduction in fetal mortality and morbidity would be small compared with improvements in general health, socio-economic status, nutrition, and disease control.

Outcome measures

Mortality and long-term morbidity

The appropriate use of electronic FHR monitors remains controversial, especially with regard to who needs monitoring and when. Initial proposals for routine monitoring in labour were accompanied by a belief that it would be an excellent method of screening for fetal hypoxia and, thereby, a means of preventing intrapartum fetal death. Some studies showed that this was the case when routine intrapartum monitoring was introduced but intrapartum stillbirth rates were high at this time. Subsequent rates of intrapartum fetal death was reduced to around 0.3 to 0.5 per 1000, but whether this was due only to the monitoring or to other associated in obstetric practices is not clear. The Dublin study (MacDonald *et al*. 1985) showed that electronic monitoring was not able to reduce intrapartum fetal mortality below this low rate, which they had already achieved by a sound practice of intermittent auscultation.

The large Dublin study did show that continuous FHR monitoring in labour decreased the rate of neonatal seizures when compared with intermittent auscultation. However, subsequent follow-up indicated that continuous monitoring did not produce any benefit with respect to rates of neurodevelopmental handicap.

Neonatal outcome

Apgar scores

The correlation between specific CTG changes and fetal condition at delivery, as measured by the Apgar score, is poor. Figure 18.1 shows the frequency of different CTG changes during the first stage of labour in an unselected population of 2566 women. Baseline rate and variability were normal in about 85 per cent of cases but decelerations occurred in almost half of all cases.

Figure 18.2 shows the CTG changes during monitored labours that ended with babies with an Apgar score of less than 7 at 1 min. A baseline rate of more than 160 bpm or less than 120 bpm was present in more than one-third of cases (36.1 per cent). Abnormal variability was seen in one-third of cases (33.6 per cent) and the rate of absent variability in this group was 21.6 per cent compared to 2.5 per cent in the total unselected population. Decelerations, especially late and combined decelerations, were seen in two-thirds of all cases. Figure 18.3 shows the CTG changes that occurred during labours that ended with babies with

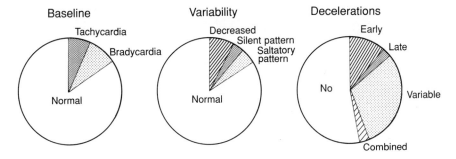

Fig. 18.1 FHR changes during the first stage of labour in an unselected population of 2566 women.

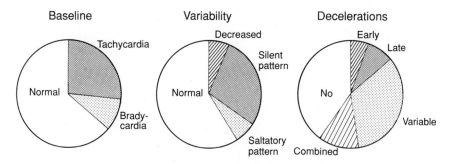

Fig. 18.2 FHR changes during the first stage of labour: newborns with an Apgar score less than 7 at 1 min.

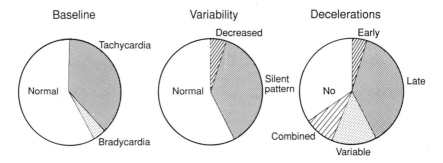

Fig. 18.3 FHR changes during the first stage of labour: newborns with an Apgar score less than 7 at 5 min.

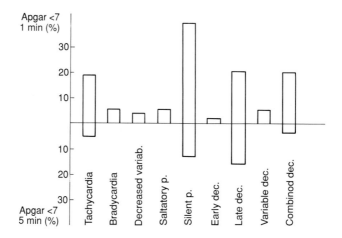

Fig. 18.4 FHR changes during the first stage of labour in an unselected population of 2566 women related to Apgar scores less than 7 at 1 and 5 min. Note, after late decelerations during the first stage of labour 21 per cent of newborns at 1 min had an Apgar score less than 7 and 16 per cent at 5 min.

Apgar scores of less than 7 at 5 min. The frequency of tachycardia in this group was even higher (38.1 per cent) and there was a similar frequency of absent variability and late decelerations.

If the parameters of the CTG are looked at in isolation it is clear that many newborns have normal Apgar scores despite changes such as tachycardia, absent variability, and late or combined decelerations during labour (Fig. 18.4). Only 19 per cent of all babies with tachycardia during labour had Apgar scores of less than 7 at 1 min and only 5 per cent had a low 5 min score. However, if tachycardia is combined with other FHR changes (Fig. 18.5), then the risk of a low Apgar score increases. There is a 15 per cent risk of an Apgar score less than 7 at 1 min with tachycardia alone, and tachycardia with high variability

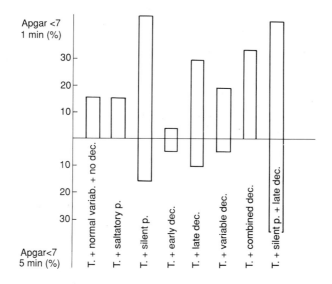

Fig. 18.5 Tachycardia (T) alone or in combination with other FHR changes related to an Apgar score less than 7 at 1 and 5 min.

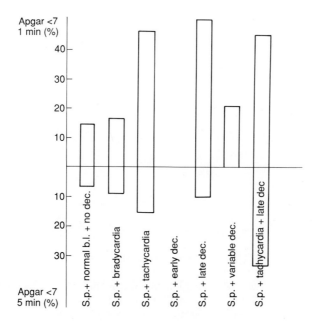

Fig. 18.6 Low or absent variability (silent pattern) alone or in combination with other FHR changes related to an Apgar score less than 7 at 1 and 5 min.

does not increase the risk of a low Apgar score. The combination of tachycardia and absent variability significantly increased the risk of low Apgar scores. If tachycardia is accompanied by late or combined decelerations, a low Apgar score is even more frequent. The highest rate of low Apgar scores (one-third less than 7 at 5 min) was seen with a combination of tachycardia, absent variability and late decelerations. In summary, this shows that uncomplicated tachycardia is not normally related to low Apgar scores, while tachycardia without variability and late/combined decelerations is strongly associated with reduced Apgar scores.

Figure 18.6 shows the combinations of low/absent variability with other CTG changes, present for more than 40 min. Absent variability and mild bradycardia (between 100 and 120 bpm) have a low frequency of reduced Apgar scores, while the opposite is true if absent variability is combined with late decelerations.

Neonatal acidaemia

Most records with ominous FHR changes are not associated with fetal acidosis in labour, birth asphyxia, or long-term sequelae. A low scalp blood pH may be found in up to 50 per cent of cases with ominous CTG changes, but in a considerable proportion of these cases the fetus has a mild to moderate

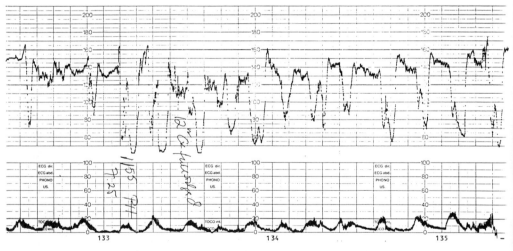

Fig. 18.7 A primipara at 38 weeks, early in the second stage of labour. Pronounced variable decelerations have been present for 2 h with a normal baseline rate and variability. Scalp blood pH was 7.32 1 h before this record. Scalp blood pH (at mark 133) was 7.25 with a normal base deficit (1.0 mmol/l) but a high P_{CO_2} (8.73 kPa). A forceps delivery was performed 7 min after this record. Apgar scores were normal. There was a respiratory acidosis in the umbilical artery with a pH of 7.11, P_{CO_2} 11 kPa, and base deficit 6.7 mmol/l.

respiratory acidosis, which is not normally associated with neonatal clinical depression and subsequent handicap. Cord compression is probably the cause in most of these cases, giving rise to pronounced variable decelerations and a respiratory acidosis due to CO_2 accumulation. If the compression is not pronounced and not of long duration the newborn is likely to have a good Apgar score (Fig. 18.7).

On the other hand, less 'dramatic' appearances on the CTG, perhaps with no decelerations at all or mild to moderate decelerations, can be associated with severe metabolic acidosis and, in some cases, long-term sequelae or fetal death. Baseline variability provides some indication of fetal reserve and may be more predictive than decelerations (Fig. 18.8 and Chapters 5 and 15).

FHR patterns suggesting fetal hypoxia may be associated with fetal acidosis and a depressed newborn if the acidosis is pronounced (cord artery pH less than 7). However, the majority of these newborns will still have a normal neonatal period and long-term outcome. Goldaber *et al.* (1991) reported that two of three term newborns had an uneventful neonatal period despite a cord artery pH below 7 at birth (8 per cent died, 10 per cent had neonatal seizures, and 20 per cent needed intensive treatment). The fetal outcome was significantly better with pH values above 7. If the cord artery pH was between 7 and 7.04, 90 per cent of the term infants had a normal neonatal outcome. In agreement with these findings, Winkler *et al.* (1991) found that only two of 23 term babies with a cord artery pH <7 had significant neonatal complications. No long-term follow-up of the infants in these studies was performed.

Asphyxia and neurodevelopmental outcome

Intrapartum asphyxia is responsible for only a minor proportion of cases of brain damage, such as cerebral palsy, according to recent long-term studies from the USA, Australia, and Sweden. In only 10–20 per cent of all children with spastic cerebral palsy was intrapartum asphyxia considered the likely cause of the brain damage. In some of these cases confounding factors were present, which made the relationship even more uncertain. In fact, it has been suggested that one explanation for an association between birth asphyxia and spastic cerebral palsy is that asphyxia is more likely to occur if there has already been damage during the antenatal period or if there is a greater predisposition to damage during labour. Thus, severe acidaemia may be a marker rather than the cause. Brain damage may already be present when labour starts and, although intervention may not improve the outcome (Fig. 18.9), it may prevent it becoming worse.

Antecedents of hypoxic ischaemic brain damage are not seen clearly during labour in a considerable proportion of cases. A recent British study (Murphy *et al.* 1990) showed that 43 of 107 cases with this neonatal outcome had a normal vaginal delivery. In almost half of these cases (43 per cent) there were no indications of fetal hypoxia using current methods of fetal monitoring.

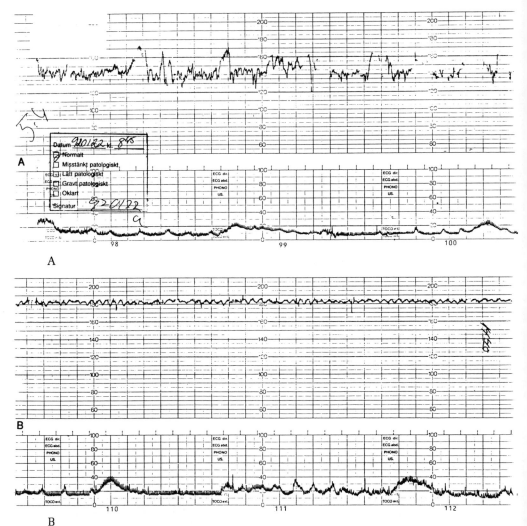

Fig. 18.8 A primipara at 35 weeks with polyhydramnios (the fetus had duodenal atresia). A, A normal CTG at 8.45 in the morning. A fetal tachycardia was noticed at ultrasound examination in the afternoon and another CTG was performed; B, sinusoid baseline with tachycardia. Computer analysis (System 8000, Oxford Sonicaid Ltd, Chichester, UK) showed reduced short-term variation at 1.3 ms (normal values above 3.8 ms, Dawes et al. 1991). The overall variation was also very low (6.5 ms); a value below 20 ms is abnormal (see Fig. 14.20). The woman had no abdominal pain or vaginal bleeding. An emergency Caesarean section was performed and the amniotic fluid was found to be blood-stained. An abruption of the placenta had occurred. The fetus was severely depressed with Apgar scores of 2, 3, and 4 at 1, 5, and 10 min, respectively. The cord artery pH was 7.21 (base deficit 10.1 mmol/l). Neonatal condition improved considerably after blood transfusion.

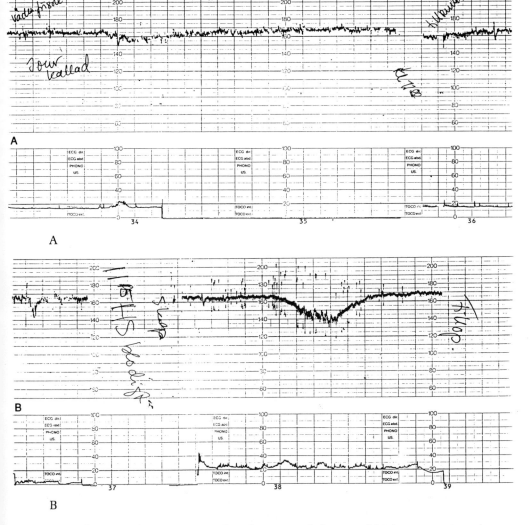

Fig. 18.9 A multipara at 38 weeks who reported absence of fetal movements and fetal hiccups for 24 h. A, The baseline tachycardia was at 180 bpm, with absent short-term variability (external recording of poor quality); B, amniotomy was performed and a scalp electrode attached to the fetal head. A poor quality record showed a deceleration. Fetal intrauterine seizures were suspected at ultrasound examination. An emergency Caesarean section was performed. The baby was severely depressed with Apgar scores of 3, 5, and 7 at 1, 5, and 10 min. Convulsions occurred immediately after birth during resuscitation. The cord artery pH was 7.28, indicating an absence of significant intrauterine asphyxia at delivery. The baby was brain-dead, presumably as a result of an intrauterine event prior to admission, and died 2 days later. The antepartum hiccups may well have been seizures *in utero*.

Acidaemia at birth, therefore, is a poor predictor of long-term, neurodevelopmental, abnormality. Many neonates, born severely acidaemic (pH less than 7.05) require no resuscitation and compensate appropriately for the acidaemia, usually within the first hour or so of life. More epidemiological work is required to obtain a fuller understanding of the relationship between acidaemia and cerebral damage. However, it seems likely that the rapid development of acidaemia during lablour, in the absence of iatrogenic factors, is indicative of poor fetoplacental reserve and the possibility that cerebral damage may already have occurred. Neonatal encephalopathy (with seizures) should only be attributed to 'birth asphyxia' if associated with evidence of fetal distress during labour, an objective measure of hypoxia at birth (low pH) and other neonatal evidence of the effects of severe hypoxia such as renal or myocardial damage. Increasing severity of encephalopathy is associated with a greater frequency of more severe neurodevelopmental handicap later in life.

Recommendations for monitoring in clinical practice

FIGO guidelines

The Federation Internationale of Gynaecologists and Obstetricians (FIGO) appointed a Subcommittee on Standards in Perinatal Medicine to draw up guidelines for the use of monitoring and interpretation of FHR patterns. The subcommittee recommended that antenatal CTG records were only made when there was a clinical indication, being appropriate in pregnancies at high risk of placental dysfunction. The value of FHR monitoring of high-risk women in labour was considered certain but there remained differences of opinion as to the true value of routine monitoring in all labours. The Admission Test was considered by many to be of value.

Other consensus reports have stated that FHR monitoring should be strongly considered in any 'at risk' patient. It is recognized that fetal hypoxia during labour may occur in any pregnancy, although it is more common in a pregnancy with risk factors for placental dysfunction. Unexpected events can arise during labour, even in low-risk labours, which subsequently require intensive surveillance. Equipment and training for FHR monitoring must be considered at any birth centre. However, routine continuous FHR monitoring is not necessary for most women and should be restricted as far as possible to avoid the associated disadvantages of the method. Antenatal risk classification should be used to identify women at increased risk of fetal hypoxia, who may be offered routine monitoring in labour. Intrapartum monitoring may then be by intermittent auscultation for low-risk women, assuming staffing levels are sufficient for this labour-intensive method.

Risk assessment

Criteria usually used for high-risk categorization include oligohydramnios, hypertension, pre-eclampsia, post-term (after 42 weeks), intrauterine growth retardation, multiple pregnancy, diabetes mellitus, rhesus isoimmunization, suspected placental abruption, preterm labour (before 37 weeks), prelabour rupture of the membranes in the preterm period, fetal malpresentation, severe medical diseases, and elderly primipara. In addition, a few situations like history of infertility, previous Caesarean section, and bad obstetric history may also be listed. If the FHR changes in such high-risk groups are compared with those in a low-risk group the differences are smaller than may be expected. Tachycardia, absent variability, and late and variable decelerations are more common in high-risk groups but there are no obvious differences for other features of the FHR pattern. As previously stated (Chapter 3) scalp blood pH values below 7.25 are found only slightly more frequently in high-risk cases than in low-risk labours. However, the real difference may be somewhat reduced by the fact that some high-risk pregnancies are delivered by elective Caesarean section and never go into labour. The large variety of perceived risks that are put together, many of which do not carry an appreciable risk of placental dysfunction, may also account for the small difference between groups distinguished by these criteria alone, especially as many cases of true placental insufficiency, with associated fetal growth retardation, remain unrecognized until birth.

Other studies have shown that risk profiles may be rather indiscriminate for intrapartum events. Hobel *et al.* (1973) studied four groups of patients in a prenatal and intrapartum risk-screening programme. Women remaining low-risk throughout pregnancy and labour (46 per cent), still contributed to 15 per cent of the total perinatal mortality and morbidity. Twenty per cent of women with a low-risk pregnancy became high-risk in labour and 33 per cent of perinatal mortality and morbidity could be attributed to this group. Women classified as high-risk in pregnancy but having a low-risk labour (18 per cent of the population) contributed 12 per cent to the perinatal mortality and morbidity figure. There are large differences between and within different high risk groups. Fetal outcome may be better in a high-risk group of women with well controlled gestational diabetes compared to a group with intrauterine growth retardation. Pregnancies with vaginal bleeding of different origin may have different risk for adverse fetal outcome.

Fetal growth retardation

Intrauterine growth retardation constitutes the major high-risk group. This diagnosis is common in pregnancies with ominous FHR changes before and during labour. In particular, cases of intrauterine growth retardation with oligohydramnios often have ominous CTG records. Occasional spontaneous contractions can provoke late decelerations. Variable decelerations on the CTG are not uncommon if oligohydramnios is severe. Also, unusual patterns, such as decelerations associated with fetal movements, may be seen. The diagnosis of fetal growth retardation attests to the risk of placental function being reduced and increases the risk of chronic fetal hypoxia.

Post-term pregnancy

Post-term pregnancies are widely considered to be at increased risk. There have been many proposals for management by fetal surveillance of these women. The risk for ominous FHR changes in labour is increased (about 15 per cent) regardless of whether expectant management, awaiting spontaneous labour, or routine induction of labour at 42 weeks is practised. In a post-term population (dated by a second trimester ultrasound scan) without other complications, after a normal CTG, and with a normal amount of liquor when tested 14 days after the estimated day of delivery, more than half went into spontaneous labour within 3 days and 90 per cent within a week (Heden *et al.* 1991). During this period of waiting for the spontaneous onset of labour, the risk of an ominous CTG is small when testing is performed on alternate days. As oligohydramnios may appear within 1–2 days, an ultrasound scan on alternate days for assessment of liquor volume may be advantageous, although neither management strategy has been shown to be better than weekly testing. The role of Doppler ultrasound in this group also remains to be determined. Others have suggested frequent oxytocin challenge tests or a biophysical profile assessment regularly in this period, confirming the lack of consensus about management.

Other risk groups

Other important high-risk groups are women with diabetes or hypertension. In one study, ominous FHR patterns and/or low scalp blood pH (less than 7.25) in labour were more common in women with diabetes compared to controls (17.4 and 10.9 per cent, respectively). Women with hypertension in pregnancy also have an increased rate of ominous FHR changes in labour (see Table 16.1).

Intermittent and continuous FHR monitoring

In general, however, risk classification profiles are not good at selecting women who require FHR monitoring in labour. Fetal surveillance in labour is not just a question of CTG or no CTG but is a question of continuous versus intermittent FHR monitoring. Whether to monitor the FHR continuously or to use intermittent monitoring should be assessed individually both at the onset and throughout the course of labour.

Admission Test

It is reasonable to obtain a short record of the continuous FHR from all women at the onset of labour (or following admission) but to restrict subsequent continuous FHR monitoring to high-risk cases. A flow chart such as the one in Figure 18.10 could be adapted according to local conditions, such as available equipment and availability of trained staff to interpret FHR records. An Admission Test could be used to exclude clear evidence of fetal hypoxia present at admission. Continuous monitoring is often indicated in cases with abnormal uterine activity, prematurity or postmaturity, vaginal bleeding, growth retarded fetus, meconium in liquor, oligohydramnios, fetal malposition, pre-eclampsia, polyhydramnios, previous Caesarean section, auscultated abnormal fetal heart tones, and maternal or fetal illnesses.

Intermittent CTG records

An alternative and increasingly popular method of intermittent monitoring in labour, applicable to many women with antenatal risk factors, is by short CTG records repeated every hour or so. The record may be as short as 10 to 15 min if the record is reactive and no decelerations are present. However, if the CTG is equivocal, as will be likely during a quiet fetal behavioural state, the record will need to be prolonged until reassurance is obtained. Women may be ambulant between intermittent CTG records but intermittent auscultation should continue during these periods. Whether intermittent short CTG records are sufficient to negate the need for intermittent auscultation between records remains to be seen.

Examples of situations where a change from intermittent auscultation to continuous FHR monitoring is indicated are the administration of an epidural block, the appearance of meconium (particularly thick — grade 3), auscultation of FHR abnormalities, commencement of an oxytocin infusion, and when labour

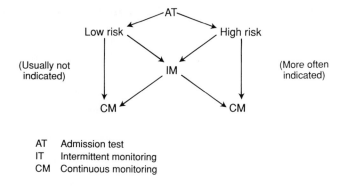

Fig. 18.10 A decision flow-chart for FHR monitoring.

is prolonged or abnormal. The risk of fetal hypoxia in labour seems to increase after about 6 h and it is reasonable to consider more careful fetal surveillance after this. Shorter intervals between CTG records of the FHR and longer periods of recording are possible alternatives to continuous monitoring. FHR changes appear predominantly at the end of the first stage of labour, which is another time to consider changing to continuous FHR monitoring. Similarly, the second stage of labour should be monitored continuously if FHR changes are auscultated. Fetal hypoxia is more common if the second stage lasts longer than 40 min and this stage should be curtailed in the presence of an abnormal FHR. The need for continuous monitoring is therefore greater in primipara, who normally have a much longer second stage.

Intermittent monitoring of the FHR, either by intermittent auscultation or intermittent use of the monitor, has many potential advantages. Women can remain ambulant, which may be of help during labour. Consumer participation in the decision as to how to monitor is becoming important for certain groups of women who feel their autonomy threatened by the impersonal adherence to labour ward policies. There is no reason for continuous FHR monitoring to interfere with childbirth in a number of alternative postures. In fact, skilful interpretation of an adequate record of the FHR (even without a toco record, which is only really necessary when there is no observer present to assess the relationship between decelerations and contractions) can be most reassuring to *accoucheur* and woman at times when intermittent auscultation would be physically difficult and would interfere with unconventional postures such as squatting.

A study of intermittent versus continuous CTG

Intermittent versus continuous CTG monitoring during the first stage of labour has recently been investigated as a randomized study (Ingemarsson *et al.*, unpublished observations). During a 19-month period, a total of 5647 low-risk women were randomized to a continuous monitoring (CM group) or intermittent (IM group) policy after an Admission Test. The FHR was recorded for 15–20 min every second hour in the IM group with stethoscopic auscultation between. Longer periods of FHR monitoring were used if an oxytocin infusion was necessary or after an epidural block was given. Telemetry was used in the CM group to match for ambulant care.

A similar rate of suspicious and ominous FHR changes were found in the two groups. Caesarean secions for fetal distress were performed in only 1.2 per cent in the CM group, indicating that a low rate can be achieved despite continuous monitoring. In the intermittently monitored group the figure was similar (1.0 per cent). Although fetal outcome was good in both groups, low Apgar scores at 1 and 5 min were more frequent in the CM group. The results indicate that intermittent monitoring in labour is a safe method for FHR monitoring in women with low risk pregnancies.

CTG scoring systems in labour

As with assessment of the antenatal CTG, there are several scoring systems for the assessment of the CTG during labour. Table 18.1 illustrates such a scoring system.

Like the Apgar score, there are five variables, each given between 0 and 2 points. The authors claim a good correlation between these scores and Apgar scores at birth. Scores between 8 and 10 would be incompatible with fetal asphyxia. Scores between 6 and 7 may indicate a compensated fetal distress according to the authors. Scores between 0 and 5 are considered abnormal, where the possibility of fetal distress is significant.

This type of scoring system has been criticized for being too complicated for daily clinical practice. An experienced interpreter might not need a scoring system for routine use but objective measures are necessary for comparative evaluations of CTG interpretation and are a requirement when untrained observers are required to distinguish pathophysiological from physiological changes.

The Krebs score has been applied to the analysis of the time taken to develop acidosis (a pH less than 7.26) when pronounced variable or late decelerations are present with reduced variability. Tracings with a score of 4 or less indicated assessment by scalp blood pH. Fetuses who had already developed acidosis were excluded from the study. The records were observed for between 30 and 240 min until delivery. During the first interval of 90–100 min the occurrence of fetal acidosis was rather infrequent but thereafter increased rapidly. Half of all cases had reached a pH value below 7.26 by 115 min, when late decelerations were present; the corresponding time for variable decelerations was 145 min and for absent variability was 185 min (Fleischer et al. 1982).

Table 18.1 Intrapartum CTG scoring system (Krebs et al. 1979a,b)

Points	0	1	2
Baseline rate (bpm)	< 100 > 180	100–119 161–180	120–160
Baseline variability			
oscillatory amplitude (bpm)	< 3	3–5 > 25	6–25
oscillatory frequency/min	< 3	3–6	> 6
Accelerations/30 min	0	periodic 1–4 sporadic	> 5
Decelerations	Late, severe variable	Mild or moderate variable Dip 0	None early

Risks of FHR monitoring

Complications

Direct fetal complications due to FHR monitoring are rare and consist mainly of small scalp infections at the site of the scalp electrode. In a few cases, a scalp abscess has developed, requiring drainage and more intensive treatment. The frequency of such infections is well below 1 per cent. Rarely, damage such as leakage of cerebrospinal fluid from the site of a scalp electrode, and trauma to the face, has been reported. Other fetal complications are scalp haematomas or arterial bleeding after scalp blood sampling.

The risk of maternal infection after internal monitoring is not considered large, although, again, there has not been much published on this. Women with prolonged labour and prolonged ruptured membranes will be exposed to many vaginal examinations and are at greater risk of intrauterine infection. However, these are also the cases where internal monitoring is most likely to have been indicated. The most important factor for infection seems to be the mode of delivery, with substantially increased rates after abdominal delivery and little risk after vaginal delivery, regardless of the type of monitoring. Invasive monitoring is probably an independent variable which contributes very little to the increased infection rate.

Increased intervention

The great disadvantage of FHR monitoring is the increased rate of intervention, which is a result of the information available from the continuous FHR record. The increasing rates of Caesarean section in many Western countries in recent years may be attributed to a lesser acceptance of risk and a greater desire for documented records (such as the CTG), which are not capable of being interpreted with sufficient precision. However, in departments with staff trained in interpretation of FHR patterns, and with access to fetal scalp blood sampling equipment, the Caesarean section rate for fetal distress can be maintained at a rate of between 1 and 2 per cent. In fact, the Caesarean section rate in Sweden declined from 12.2 per cent in 1983 to 11 per cent in 1989. During the same period the rate of abdominal deliveries for fetal distress was reduced from 3.9 to 2 per cent ($p < 0.001$) indicating that the Caesarean section rate can be controlled on a national basis.

Medicalization of childbirth

Evidence that routine monitoring will reduce fetal mortality and morbidity below present rates is lacking, although experienced users will give evidence of benefit in individual cases. FHR changes are difficult to classify and there are well documented differences of opinion about interpretation. Frequent FHR changes cause anxiety and mislead clinicians, particularly inexperienced ones, resulting in high rates of intervention. Therefore, with routine use of electronic fetal monitors, there is a considerable risk of an increased Caesarean section rate for suspected fetal distress, as shown by several randomized studies. Prentice and Lind (1987) suggest that for low-risk mothers there is a good case for returning to the traditional method of intermittent auscultation.

In addition, concern about the 'medicalization' of childbirth has been expressed by various consumer groups. There is concern that monitors are used just because they are available rather than because they are needed. A relative shortage of midwives may also be a contributory factor to their continuing use in apparently low-risk cases.

Summary

It should be stressed that the FHR is only one parameter used for monitoring fetal condition. Other clinical parameters need to be considered at all times in order to obtain good results. Pathological changes on a CTG should confirm a diagnosis that is suspected and for which the test is being performed. The performance of a test without knowing the meaning of the possible results has been one of the problems with the use of the CTG, especially during pregnancy.

Although there are many FHR changes during labour, even in an unselected population, only between 6 and 8 per cent are ominous. Trained staff should have few problems with recognition and management. In such cases, determination of fetal pH in scalp blood will confirm or refute a diagnosis of significant fetal hypoxia. FHR changes in some circumstances indicate prompt delivery rather than delaying delivery by fetal scalp blood sampling.

Other than in a few acute obstetric situations, when severe fetal bradycardia may demand immediate action, there is time for an analysis of the circumstances associated with FHR changes. The history and clinical examination are of paramount importance and should direct the search for an explanation. Knowledge of fetal physiology is also a prerequisite, and it is clear that interpretation of CTG patterns cannot be learnt without understanding the potential pathophysiological mechanisms that influence control of the FHR.

Bibliography

General

Balen, A. H. and Smith, J. H. (1992). *The CTG in practice*. Churchill Livingstone, London.
Cruikshank, D. P. (ed.) (1982) *Antepartum fetal surveillance, Clinical Obstetrics and Gynecology*, Vol. 4. W. B. Saunders, Philadelphia.
FIGO News. (1987). Guidelines for the use of fetal monitoring. *International Journal of Gynaecology and Obstetrics*, **25**, 159–67.
Fischer, W. M., Berg, B., Brandt, H., Ekut, W. P., Gennser, G., Halberstadt, J., Heep, A., Huch, A., Huch, R., Junge, H. D., Leodolter, S., Philipp, K., Rooth, G., Ruttgers, H., and Stude, I. (1981). *Kardiotokographie*. Georg Thieme Verlag (3. Auflage), Stuttgart.
Freeman, R. K. and Garite, T. J. (1981). *Fetal heart rate monitoring*. Williams & Wilkins, Baltimore.
Gibb, D. and Arulkumaran, S. (1992). *Fetal monitoring in practice*. Butterworth-Heinemann Ltd, Oxford.
Ingemarsson, E. (1981). Routine electronic monitoring during labor. *Acta Obstetricia et Gynecologica Scandinavica*, **Suppl. 99**.
Parer, J. T. (1983). *Handbook of fetal heart rate monitoring*. W. B. Saunders, Philadelphia.
Paul, R., Petrie, R., Rabello, Y. A., and Mueller, E. A. (1979). *Fetal intensive care*. Corometrics Medical Systems Inc., Wallingford, Oxfordshire.
Petrie, R. H. (ed.) (1986) Intrapartum fetal monitoring. *Clinical Obstetrics and Gynecology*, Vol. 29. W. B. Saunders, Philadelphia.
Quilligan, E. J. (ed.) (1979). Update on fetal monitoring. *Clinics in obstetrics and gynaecology*, Vol. 6. W. B. Saunders, Philadelphia.
Schifrin, B. S. (1990). *Exercises in fetal monitoring*. Mosby Year Book, St Louis.
Spencer, J. A. D. (1990). Antepartum cardiotocography. In *Modern antenatal care of the fetus* (ed. G. V. P. Chamberlain), pp. 163–88. Blackwell Scientific Publications, Oxford.
Spencer, J. A. D. (1991). *Fetal monitoring*. Oxford University Press, Oxford.

Chapter 1

Banta, H. D. and Thacker, S. B. (1979). Assessing the costs and benefits of electronic fetal monitoring. *Obstetrics and Gynecology Survey*, **34**, 627–42.
Hobbins, J. C., Freeman, R., and Queenan, J. T. (1979). The fetal monitoring debate. *Obstetrics and Gynecology*, **54**, 103–9
Ingemarsson, E., Ingemarsson, I., Solum, T., and Westgren, M. (1980). A one-year study of routine fetal monitoring during the first stage of labour. *Acta Obstetricia et Gynecologica Scandinavica*, **59**, 297–300.
Schifrin, B. S. and Dame, L. (1972). Fetal heart rate patterns. Prediction of Apgar score. *Journal of the American Medical Association*, **219**, 1322–5.

Chapter 2

Amato, J. C. (1983). Fetal heart rate monitoring. *American Journal of Obstetrics and Gynecology*, **147**, 967-9.

Dawes, D. S., Visser, G. H. A., Goodman, J. D. S., and Redman C. W. G. (1981). Numerical analysis of the human fetal heart rate: The quality of ultrasound records. *American Journal of Obstetrics and Gynecology*, **141**, 43-52.

Divon, M. Y., Torres, F. P., Yeh, S.-Y., and Paul, R. H. (1985). Autocorrelation techniques in fetal monitoring. *American Journal of Obstetrics and Gynecology*, **151**, 2-6.

Solum, T. (1980). A comparison of the three methods for external fetal cardiography. *Acta Obstetricia et Gynecologica Scandinavica*, **59**, 123-6.

Solum, T., Ingemarsson, I., and Nygren, A. (1980). The accuracy of abdominal ECG for fetal electronic monitoring. *Journal of Perinatal Medicine*, **8**, 142-9.

Solum, T., Ingemarsson, I., and Nygren, A. (1981). The accuracy of ultrasonic fetal cardiography. *Journal of Perinatal Medicine*, **9**, 54-62.

Spencer, J. A. D. (1987). The quality of ultrasound telemetry fetal heart rate records. *Journal of Obstetrics and Gynaecology*, **8**, 12-15.

Spencer, J. A. D., Belcher, R., and Dawes, G. S. (1987). The influence of signal loss on the comparison between computer analyses of the fetal heart rate in labour using pulsed Doppler ultrasound (with autocorrelation) and simultaneous scalp electrocardiogram. *European Journal of Obstetrics, Gynecology and Reproductive Biology*, **25**, 29-34.

Chapter 3

Beard, R. W. and Morris, E. D. (1965). Foetal and maternal acid-base balance during normal labour. *The Journal of Obstetrics and Gynaecology of the British Commonwealth*, **72**, 496-506.

Clark, S. and Paul, R. H. (1985). Intrapartum fetal surveillance: The role of fetal scalp blood sampling. *American Journal of Obstetrics and Gynecology*, **153**, 717-20.

Eskes, T. K. A. B., Jongsma, H. W., and Houx, P. C. W. (1983). Percentiles for gas values in human umbilical cord blood. *European Journal of Obstetrics, Gynecology and Reproductive Biology*, **14**, 341-6.

Humphrey, M. D., Chang, A., Wood, E. C., Morgan, S., and Hounslow, D. (1974). A decrease in fetal pH during the second stage of labour, when conducted in the dorsal position. *The Journal of Obstetrics and Gynaecology of the British Commonwealth*, **81**, 600-2.

Ingemarsson, I. and Arulkumaran, S. (1986). Fetal acid-base balance in low-risk patients in labor. *American Journal of Obstetrics and Gynecology*, **155**, 66-9.

Josten, B. E., Johnson, R. B., and Nelson, J. P. (1987). Umbilical cord blood pH and Apgar scores as an index of neonatal health. *American Journal of Obstetrics and Gynecology*, **157**, 843-8.

Kerenyi, T. D., Falk, S., Mettel, R. D., and Walker, B. (1970). Acid-base balance and oxygen saturation of fetal scalp blood during normal and abnormal labors. *Obstetrics and Gynecology*, **36**, 398-404.

MacLachlan, N. A., Spencer, J. A. D., Harding, K., and Arulkumaran, S. (1992). Fetal acidaemia, the cardiotocograph and the T/QRS ratio of the fetal ECG in labour. *British Journal of Obstetrics and Gynaecology*, **99**, 26-31.

Pearson, J. P. and Davies, P. (1974). The effect of continuous lumbar epidural analgesia upon fetal acid-base status during the second stage of labour. *British Journal of Obstetrics and Gynaecology*, **81**, 975-9.

Roemer, V. M., Harms, K., Buess, H., and Horvath, T. J. (1976). Response of fetal acid-base balance to duration of second stage of labour. *International Journal of Gynaecology and Obstetrics*, **14**, 455-71.

Rooth, G. (1964). Early detection and prevention of foetal acidosis. *Lancet*, **1**, 290-3.

Saling, E. (1981). Fetal scalp blood analysis. *Journal of Perinatal Medicine*, 165-77.

Spencer, J. A. D., Robson, S. C., and Farkas, A. (1991). Spontaneous correction of severe acidaemia after birth. *Early Human Development*, **26**, 235-6.

Thorp, J. A., Sampson, J. E., Parisi, V. M., and Creasy, R. K. (1989). Routine umbilical cord blood gas determinations? *American Journal of Obstetrics and Gynecology*, **161**, 600-5.

Wheble, A. M., Gillmer, M. D. G., Spencer J. A. D., and Sykes, G. S. (1989). Changes in fetal monitoring practice in the UK: 1977-1984. *British Journal of Obstetrics and Gynaecology*, **96**, 1140-7.

Yeomans, E. R., Hauth, J. C., Gilstrap, L. C., and Strickland, D. M. (1985). Umbilical cord pH, $P\text{CO}_2$, and bicarbonate following uncomplicated term vaginal deliveries. *American Journal of Obstetrics and Gynecology*, **151**, 798-800.

Yudkin, P. L., Johnson, P., and Redman, C. W. G. (1987). Obstetric factors associated with cord blood gas values at birth. *European Journal of Obstetrics, Gynecology and Reproductive Biology*, **24**, 167-76.

Chapter 4

Arulkumaran, S. and Ingemarsson, I. (1989). Methods for measurement of uterine activity in labour. *International Journal of Feto-Maternal Medicine*, **2**, 23-8.

Arulkumaran, S., Ingemarsson, I., and Ratnam, S. S. (1987). Oxytocin titration to achieve preset active contraction area values does not improve the outcome of induced labour. *British Journal of Obstetrics and Gynaecology*, **94**, 272-8.

Arulkumaran, S., Ingemarsson, I., Gibb, D. M. F., and Ratnam, S. S. (1988). Uterine activity in spontaneous labour with breech presentation. *Australian and New Zealand Journal of Obstetrics and Gynaecology*, **28**, 275-8.

Arulkumaran, S., Ingemarsson, I., Gibb, D. M. F., Kitchener, C. H., and Ratnam, S. S. (1989). Uterine activity during spontaneous labour after previous lower segment caesaraean section. *British Journal of Obstetrics and Gynaecology*, **96**, 933-8.

Baumgarten, K. (1981). The advantages and risks of feto-maternal monitoring. *Journal of Perinatal Medicine*, **9**, 257-74.

Caldeyro-Barcia, R. and Poseiro, J. J. (1960). Physiology of uterine contractions. *Clinical Obstetrics and Gynecology*, **3**, 386-408.

Gibb, D. M. F., Arulkumaran, S., Lun, K. C., and Ratnam, S. S. (1984). Characteristics of uterine activity in nulliparous labour. *British Journal of Obstetrics and Gynaecology*, **91**, 220–7.
Hon, E. H. and Paul, R. H. (1973). Quantitation of uterine activity. *Obstetrics and Gynecology*, **42**, 368–70.
Phillips, G. F. and Calder, A. A. (1987). Units for evaluation of uterine contractility. *British Journal of Obstetrics and Gynaecology*, **94**, 236–41.
Seitchik, J. and Chatkoff, M. L. (1976). Oxytocin-induced uterine hypercontractility pressure wave forms. *Obstetrics and Gynecology*, **48**, 436–41.
Steer, P. J., Carter, M. C., Gordon, A. J., and Beard, R. W. (1978). The use of catheter-tip pressure transducers for the measurement of intra-uterine pressure in labour. *British Journal of Obstetrics and Gynaecology*, **85**, 561–6.
Steer, P. J., Carter, M. C., and Beard, R. W. (1984). Normal levels of active contraction area in spontaneous labour. *British Journal of Obstetrics and Gynaecology*, **91**, 211–19.

Chapter 5

Dalton, K. J., Dawes, G. S., and Patrick, J. E. (1983). The autonomic nervous system and fetal heart rate variability. *American Journal of Obstetrics and Gynecology*, **146**, 456–62.
Dawes, G. S., Houghton, C. R. S., Redman, C. W. G., and Visser, G. H. A. (1982). Pattern of the normal human fetal heart rate. *British Journal of Obstetrics and Gynaecology*, **89**, 276–84.
Escarcena, L., McKinney, R. D., and Depp, R. (1979). Fetal baseline heart rate variability estimation. *American Journal of Obstetrics and Gynecology*, **135**, 615–21.
van Geijn, H. P., Jongsma, H. W., de Haan, J., Eskes, T. K. A. B., and Prechtl, H. F. R. (1980). Heart rate as an indicator of the behavioral state. *American Journal of Obstetrics and Gynecology*, **136**, 1061–6.
Hammacher, K., Huter, K. A., Bokelmann, J., and Werners, P. H. (1968). Foetal heart frequency and perinatal condition of the foetus and newborn. *Gynaecologica*, **166**, 349–60.
Modanlou, H. and Freeman, R. (1982). Sinusoidal fetal heart rate pattern: Its definition and clinical significance. *American Journal of Obstetrics and Gynecology*, **142**, 1033–8.
Murata, Y., Miyake, Y., Yamamoto, T., Higuchi, M., Hesser, J., Ibara, S., Bessho, T., and Tynor, J. G. (1953). Experimentally produced sinusoidal fetal heart rate pattern in the chronically instrumented fetal lamb. *American Journal of Obstetrics and Gynecology*, **153**, 693–702.
Paul, R. H., Suidan, A. K., Yeh, S-Y., Schifrin, B., and Hon, E. H. (1975). Clinical fetal monitoring. VII. The evaluation and significance of intrapartum baseline FHR variability. *American Journal of Obstetrics and Gynecology*, **123**, 206–10.
Pello, L. C., Rosevear, S. K., Dawes, G. S., Moulden, M., and Redman, C. W. G. (1991). Computerized fetal heart rate analysis in labor. *Obstetrics and Gynecology*, **78**, 602–10.
Sorokin, Y., and Dierker, L. J. (1982). Fetal movement. *Clinical Obstetrics and Gynecology*, **25**, 719–34.

Spencer, J. A. D. (1989). Fetal heart rate variability. In *Progress in Obstetrics and Gynaecology* (ed. J. Studd), pp. 103-22. Churchill Livingstone, London.

Spencer, J. A. D. and Johnson, P. (1986). Fetal heart rate variability changes and fetal behavioural cycles during labour. *British Journal of Obstetrics and Gynaecology*, **93**, 314-21.

Timor-Tritsh, I. E., Dierker, L. J., Hertz, R. Deagan, N., and Rosen, M. G. (1978). Studies of antepartum behavioral state in the human fetus at term. *American Journal of Obstetrics and Gynecology*, **132**, 524-8.

Visser, G. H. A., Goodman, J. D. S., Levine, D. H., and Dawes G. S. (1982). Diurnal and other cyclic variations in human fetal heart rate near term. *American Journal of Obstetrics and Gynecology*, **142**, 535-44.

Visser, G. H. A., Zeelenberg, H. J., de Vries, J. I. P., and Dawes, G. S. (1983). External physical stimulation of the human fetus during episodes of low heart rate variation. *American Journal of Obstetrics and Gynecology*, **145**, 579-84.

Wheeler, T. and Murrils, A. (1978). Patterns of fetal heart rate during normal pregnancy. *British Journal of Obstetrics and Gynaecology*, **85**, 18-27.

Young, B. K., Kats, M., and Wilson, S. J. (1980). Sinusoidal fetal heart rate. I. Clinical significance. *American Journal of Obstetrics and Gynecology*, **136**, 587-93.

Chapter 6

Beall, M. H. and Paul, R. H. (1986). Artifacts, blocks, and arrhythmias: Confusing nonclassical heart rate tracings. *Clinical Obstetrics and Gynecology*, **29**, 83-103.

Beard, R. W., Filshie, G. M., Knight, C. A., and Roberts, G. M. (1971). The significance of the changes in the continuous fetal heart rate in the first stage of labour. *Journal of Obstetrics and Gynaecology of the British Commonwealth*, **78**, 865-81.

Cohen, W. R. and Yeh, S.-Y. (1986). The abnormal fetal heart rate baseline. *Clinical Obstetrics and Gynecology*, **29**, 73-82.

Crawford, D., Chapman, M., and Allan, L. (1985). The assessment of persistent bradycardia in prenatal life. *British Journal of Obstetrics and Gynaecology*, **92**, 941-4.

Fusi, L., Maresh, M. J., Steer, P. J., and Beard, R. W. (1989). Maternal pyrexia associated with the use of epidural analgesia in labour. *Lancet*, **i**, 1250-2.

Patrick, J., Campbell, K., Carmichael, L., and Probert, C. (1982). Influence of maternal heart rate and gross fetal body movements on the daily pattern of fetal heart rate near term. *American Journal of Obstetrics and Gynecology*, **144**, 533-8.

Spencer, J. A. D., Deans, A., Nicolaidis, P., and Arulkumaran S. (1991). Fetal heart rate response to vibro-acoustic stimulation during low and high heart rate variability episodes in late pregnancy. *American Journal of Obstetrics and Gynecology*, **165**, 86-96.

Young, B. K. and Weinstein, H. M. (1976). Moderate fetal bradycardia. *American Journal of Obstetrics and Gynecology*, **126**, 271-5.

Young, B. K., Katz, M., Klein, S. A., and Silverman, F. (1979). Fetal blood and tissue pH with moderate bradycardia. *American Journal of Obstetrics and Gynecology*, **135**, 45-7.

Chapter 7

Krebs, H.-B., Petres, R. E., Dunn, L. J., Jordaan, H. V. F., and Segreti, A. (1979). Intrapartum fetal heart rate monitoring. I. Classification and prognosis of fetal heart rate patterns. *American Journal of Obstetrics and Gynecology*, **133**, 762–72.
Krebs, H.-B., Petres, R. E., Dunn, L. J., and Smith, P. J. (1982). Intrapartum fetal heart rate monitoring. VI. Prognostic significance of accelerations. *American Journal of Obstetrics and Gynecology*, **142**, 297–305.
Lee, C. Y., DiLoreto, P. C., and O'Lane, J. M. (1975). A study of fetal heart rate acceleration patterns. *Obstetrics and Gynecology*, **45**, 142–6.
Rayburn, W. F., Duhring, J. L., and Donaldson, M. (1978). A study of fetal acceleration tests. *American Journal of Obstetrics and Gynecology*, **132**, 33–5.
Sadovsky, E., Rabinowitz, R., Freeman, A., and Yarkoni, S. (1984). The relationship between fetal heart rate accelerations, fetal movements, and uterine contractions. *American Journal of Obstetrics and Gynecology*, **149**, 187–9.
Spencer, J. A. D. (1991). Predictive value of a fetal heart rate acceleration at the time of fetal blood sampling in labour. *Journal of Perinatal Medicine*, **19**, 207–15.
Wittmann, B. K., Davison, B. M., Lyons, E., Frohlich, J., and Towell, M. E. (1979). Real-time ultrasound observations of fetal activity in labour. *British Journal of Obstetrics and Gynaecology*, **86**, 278–81.
Zimmer, E. Z. and Vadasz, A. (1989). Influence of the fetal scalp electrode stimulation test on fetal heart rate and body movements in quiet and active behavioral states during labour. *American Journal of Perinatology*, **6**, 24–9.
Zimmer, E. Z., Divon, M. Y., and Vadasz, A. (1987). The relationship between uterine contractions, fetal movements and fetal heart rate patterns in the active phase of labour. *European Journal of Obstetrics, Gynecology and Reproductive Biology*, **25**, 89–95.

Chapter 8

Cibils, L. A. (1975). Clinical significance of fetal heart rate patterns during labor. II. Late decelerations. *American Journal of Obstetrics and Gynecology*, **123**, 473–94.
Cibils, L. A. (1980). Clinical significance of fetal heart rate patterns during labor. VI. Early decelerations. *American Journal of Obstetrics and Gynecology*, **136**, 392–8.
Caldeyro-Barcia, R., Mendez-Bauer, C., Poseiro, J. J., Escarcena, L. A., Pose, S. V., Bieniarz, J., Arat, I., Gulin, L., and Althabe, O. (1966). Control of human fetal heart rate during labour. In *The heart and circulation in the newborn and infant*, (ed. D. E. Cassels), pp. 7–36. Grune and Stratton, New York.
Harris, J. L., Krueger, T. R., and Parer, J. T. (1982). Mechanisms of late decelerations of the fetal heart rate during hypoxia. *American Journal of Obstetrics and Gynecology*, **144**, 491–6.
Hon, E. H. (1968). *An atlas of fetal heart rate patterns*. Harty Press, New Haven.
James, L. S., Morishima, H. O., Daniel, S. S., Bowe, E. T., Cohen, H., and Niemann, W. H. (1972). Mechanism of late deceleration of the fetal heart rate. *American Journal of Obstetrics and Gynecology*, **113**, 578–82.

Kubli, F. W., Hon, E. H., Khazin, A. F., and Takemura, H. (1969). Observations on heart rate and pH in the human fetus during labor. *American Journal of Obstetrics and Gynecology*, **104**, 1190-206.

Myers, R. E., Mueller-Heubach, E., and Adamsons, K. (1973). Predictability of the state of fetal oxygenation from a quantitative analysis of the components of late deceleration. *American Journal of Obstetrics and Gynecology*, **115**, 1083-94.

Parer, J. T., Krueger, T. R., and Harris, J. L. (1980). Fetal oxygen consumption and mechanisms of heart rate response during artificially produced late decelerations of fetal heart rate in sheep. *American Journal of Obstetrics and Gynecology*, **136**, 478-82.

Thomas, G. (1975). The aetiology, characteristics and diagnostic relevance of late deceleration patterns in routine obstetric practice. *British Journal of Obstetrics and Gynaecology*, **82**, 121-5.

Chapter 9

Benedett, T. J., Lowensohn, R. J., and Truscott, A. M. (1980). Face presentation at term. *Obstetrics and Gynecology*, **55**, 199-202.

Cibils, L. A. (1978). Clinical significance of fetal heart rate patterns during labor. V. Variable decelerations. *American Journal of Obstetrics and Gynecology*, **132**, 791-805.

Eilen, B., Fleischer, A., Schulman, H., and Jagani, N. (1984). Fetal acidosis and the abnormal fetal heart rate tracing: The term breech fetus. *Obstetrics and Gynecology*, **63**, 233-6.

Gaziano, E. P. (1979). A study of variable decelerations in association with other heart rate patterns during monitored labor. *American Journal of Obstetrics and Gynecology*, **135**, 360-3.

Goodlin, R. C. and Lowe, E. W. (1974). A functional umbilical cord occlusion heart rate pattern. The significance of overshoot. *Obstetrics and Gynecology*, **43**, 22-30.

Ingemarsson, E., Ingemarsson, I., Solum, T., and Westgren, M. (1990). Influence of occiput posterior position on the fetal heart rate pattern. *Obstetrics and Gynecology*, **55**, 301-4.

Ingemarsson, I., Arulkumaran, S., and Westgren, M. (1990). Breech delivery—management and long-term outcome. In *Obstetrical events and early-midchildhood development* (ed. N. Tejani), pp. 143-59. C. R. C. Press Inc., Boca Raton, Florida.

James, L. S., Yeh, S.-Y, Morishima, H. O., Daniel, S. S., Caritis, S. N., Niemann, W. H., and Iudyk, L. (1976). Umbilical vein occlusion and transient acceleration of the fetal heart rate. Experimental observations in subhuman primates. *American Journal of Obstetrics and Gynecology*, **126**, 276-83.

Krebs, H.-B., Petres, R. E., and Dunn, L. J. (1983). Intrapartum fetal heart rate monitoring. VIII. Atypical variable decelerations. *American Journal of Obstetrics and Gynecology*, **145**, 297-305.

Kubli, F. W., Hon, E. H., Khazin, A. F., and Takemura, H. (1969). Observations on heart rate and pH in the human fetus during labor. *American Journal of Obstetrics and Gynecology*, **104**, 1190-206.

Miyazaki, F. S. and Taylor, N. A. (1983). Saline amnioinfusion for relief of variable or prolonged decelerations. *American Journal of Obstetrics and Gynecology*, **146**, 670-8.

Mueller-Heubach, E. and Batelli, A. F. (1982). Variable heart rate decelerations and transcutaneous PO_2 during umbilical cord occlusion in fetal monkeys. *American Journal of Obstetrics and Gynecology*, **144**, 796-802.

Nordstrom, L., Ingemarsson, E., Ingemarsson, I., and Westgren, M. (1989). Pronounced variable decelerations and its relation to breech presentation. *International Journal of Feto-Maternal Medicine*, **2**, 89-92.

Rayburn, W. F., Beyhen, A., and Brinkman, D. L. (1981). Umbilical cord length and intrapartum complications. *Obstetrics and Gynecology*, **57**, 450-2.

Sokol, R. J., Roux, J. F., and McCarthy, S. (1975). Computer diagnosis of labor progression. VI. Fetal stress and labor in the occipitoposterior position. *American Journal of Obstetrics and Gynecology*, **122**, 253-60.

Tejani, N. A., Mann, L. I., Sanghavi, M., Bhakthavathsalan, A., and Weiss, R. R. (1977). The association of umbilical cord complications and variable decelerations with acid-base findings. *Obstetrics and Gynecology*, **49**, 159-62.

Wheeler, T. and Greene, K. (1975). Fetal heart rate monitoring during breech labour. *British Journal of Obstetrics and Gynaecology*, **82**, 208-14.

Yeh, M.-N., Morishima, H. O., Niemann, W. H., and James, L. S. (1975). Myocardial conduction defects in association with compression of the umbilical cord. *American Journal of Obstetrics and Gynecology*, **121**, 951-7.

Young, B. K., Katz, M., and Wilson, S. J. (1980). Fetal blood and tissue pH with variable deceleration patterns. *Obstetrics and Gynecology*, **56**, 170-5.

Chapter 10

Ingemarsson, E., Ingemarsson, I., and Westgren, M. (1981). Combined decelerations — Clinical significance and relation to uterine activity. *Obstetrics and Gynecology*, **58**, 35-9.

O'Gureck, J. E., Roux, J. F., and Neuman, M. R. (1972). A practical classification of fetal heart rate patterns. *Obstetrics and Gynecology*, **40**, 356-61.

Chapter 11

Ingemarsson, I., Arulkumaran, S., and Ratnam, S. S. (1985). Single injection of terbutaline in term labor. I. Effect on fetal pH in cases with prolonged bradycardia. *American Journal of Obstetrics and Gynecology*, **153**, 859-64.

Ingemarsson, I., Arulkumaran, S., and Ratnam, S. S. (1985). Single injection of terbutaline in term labor. II. Effect on uterine activity. *American Journal of Obstetrics and Gynecology*, **153**, 865-9.

Kinsella, S. M., Lee, A., and Spencer, J. A. D. (1990). Maternal and fetal effects of the supine and pelvic tilt positions in late pregnancy. *European Journal of Obstetrics, Gynecology and Reproductive Biology*, **36**, 11-17.

Kinsella, S. M., Whitwam, J. G., and Spencer, J. A. D. (1990). Aortic compression by the uterus: Identification with the Finapres digital arterial pressure instrument. *British Journal of Obstetrics and Gynaecology*, **97**, 700–5.

Langer, O. and Sonnendecker, E. W. W. (1982). Characteristics and management of intrapartum prolonged fetal bradycardia. *British Journal of Obstetrics and Gynaecology*, **89**, 904–12.

Pent, D. (1979). Vasa previa. *American Journal of Obstetrics and Gynecology*, **134**, 151–5.

Schneider, E. P. and Tropper, P. J. (1986). The variable deceleration, prolonged deceleration and sinusoidal fetal heart rate. *Clinical Obstetrics and Gynecology*, **29**, 64–72.

Tejani, N., Mann, L., Bhakthanathsalan, A., and Weiss, R. (1975). Prolonged fetal bradycardia with recovery—its significance and outcome. *American Journal of Obstetrics and Gynecology*, **122**, 975–8.

Chapter 12

Derham, R. J., Crowhurst, J., and Crowther, C. (1991). The second stage of labour: durational dilemmas. *Australia and New Zealand Journal of Obstetrics and Gynaecology*, **31**, 1–31.

Fischer, W. M., Fendel, M. D., and Schultze-Mosgan, H. (1975). Fetal heart rate patterns (FHRP) in the second stage of labour and the perinatal outcome. In *Perinatal medicine* (ed. Z. K. Stembera, K. Polacek, V. Sabata), pp. 54–5. Georg Thieme Verlag, Stuttgart.

Gaziano, E. P., Freeman, D. W., and Bendel, R. P. (1980). FHR variability and other heart rate observations during second stage labour. *Obstetrics and Gynecology*, **56**, 42–7.

Gilstrap, L. C., Hauth, J. C., Hankins, G. D. V., and Beck, A. W. (1987). Second-stage fetal heart rate abnormalities and type of neonatal acidaemia. *Obstetrics and Gynecology*, **70**, 191–5.

Herbert, C. and Boehm, F. (1981). Prolonged end-stage fetal heart rate deceleration: A reanalysis. *Obstetrics and Gynecology*, **57**, 589–93.

Katz, M., Shani, N., Meizner, I., and Insler, V. (1982). Is end-stage deceleration of the fetal heart ominous? *British Journal of Obstetrics and Gynaecology*, **89**, 186–9.

Katz, M., Lunenfeld, E., Meizner, I., Bashan, N., and Gross J. (1987). The effect of the duration of the second stage of labour on the acid–base state of the fetus. *British Journal of Obstetrics and Gynaecology*, **94**, 425–30.

Krebs, H.-B., Petres, R. E., and Dunn, L. J. (1981). Intrapartum fetal heart rate monitoring. V. Fetal heart rate patterns in the second stage of labor. *American Journal of Obstetrics and Gynecology*, **140**, 415–39.

Piquard, F., Hsiung, R., Mettauer, M., Schaefer, A., Haberey, P., and Dellenbach, P. (1988). The validity of fetal heart rate monitoring during the second stage of labour. *Obstetrics and Gynecology*, **72**, 746–51.

Chapter 13

Bishop, E. (1981). Fetal acceleration test. *American Journal of Obstetrics and Gynecology*, **141**, 905–9.

Bowes, W. A., Gabbe, S. G., and Bowes, C. (1980). Fetal heart rate monitoring in premature infants weighing 1500 grams or less. *American Journal of Obstetrics and Gynecology*, **137**, 791–6.

Braithwaite, N. D. J., Milligan, J. E., and Shennan, A. T. (1986). Fetal heart rate monitoring and neonatal mortality in the very preterm infant. *American Journal of Obstetrics and Gynecology*, **154**, 250–4.

Devoe, L. D. (1982). Antepartum fetal heart rate testing in preterm pregnancy. *Obstetrics and Gynecology*, **60**, 431–6.

Dierker, L. J., Pillay, S. K., Sorokin, Y., and Rosen, M. G. (1982). Active and quiet periods in the preterm and term fetus. *Obstetrics and Gynecology*, **60**, 65–70.

Holmquist, P., Svenningsen, N., Westgren, M., and Ingemarsson, I. (1983). Fetal pH and postnatal adaptation in preterm vaginal deliveries. *Journal of Perinatal Medicine*, **11**, 74–84.

Holmquist, P., Svenningsen, N. W., and Ingemarsson, I. (1984). Neurodevelopmental outcome and electronic fetal heart rate monitoring in a neonatal intensive care population. *Acta Obstetricia et Gynecologica Scandinavica*, **63**, 527–32.

Ibarra-Polo, A. A., Guiloff, E., and Gomez-Rogers, C. (1972). Fetal heart rate throughout pregnancy. *American Journal of Obstetrics and Gynecology*, **113**, 814–18.

Malcus, P., Svenningsen, N., and Westgren, M. (1986). Reactivity of non-stress tests and its relationship to outcome in infants born prior to the 33rd week of gestation. *Acta Obstetricia et Gynecologica Scandinavica*, **65**, 835–8.

Westgren, M., Holmquist, P., Svenningsen, N., and Ingemarsson, I. (1981). Intrapartal fetal monitoring in preterm deliveries. *Obstetrics and Gynecology*, **60**, 99–106.

Westgren, M., Ingemarsson, I., Ahlstrom, H., Lindroth, M., and Svenningsen, N. (1982). Delivery and long-term outcome of very low birthweight infants. *Acta Obstetricia et Gynecologica Scandinavica*, **61**, 25–30.

Westgren, M., Holmquist, P., Ingemarsson, I., and Svenningsen, N. W. (1984). Intrapartum fetal acidosis in preterm infants. A prospective study in regard to fetal monitoring and long-term morbidity. *Obstetrics and Gynecology*, **63**, 355–9.

Zanini, B., Paul, R. H., and Huey, J. R. (1980). Intrapartum fetal heart rate: Correlation with scalp pH in the preterm fetus. *American Journal of Obstetrics and Gynecology*, **136**, 43–7.

Chapter 14

Borgatta, L., Shrout, P. E., and Divon, M. Y. (1988). Reliability and reproducibility of nonstress test readings. *American Journal of Obstetrics and Gynecology*, **159**, 554–8.

Brown, R. and Patrick, J. (1981). The nonstress test: How long is enough? *American Journal of Obstetrics and Gynecology*, **141**, 646–51.

Collea, J. V. and Holls, W. M. (1982). The contraction stress test. *Clinical Obstetrics and Gynecology*, **25**, 707–17.

Dawes, G. S., Moulden, M. and Redman, C. W. G. (1991). The advantages of computerized heart rate analysis. *Journal of Perinatal Medicine*, **19**, 39–45.

Bibliography

Dawes, G. S., Moulden, M. and Redman, C. W. G. (1991). System 8000: Computerized antenatal FHR analysis. *Journal of Perinatal Medicine*, **19**, 47–51.

Ershine, R. Z. A., Ritchie, J. W. K., Zaltz, A., and Tice, T. (1986). Failure of nonstress test and Doppler-assessed umbilical arterial blood flow to detect imminent intrauterine death. *American Journal of Obstetrics and Gynecology*, **154**, 1109–10.

Evertson, L., Gauthier, R. J., Schifrin, B. S., and Paul, R. H. (1979). Antepartum fetal heart rate testing. I. Evolution of the nonstress test. *American Journal of Obstetrics and Gynecology*, **133**, 29–33.

Flynn, A. M., Kelly, J., Matthews, K., O'Conor, M., and Viegas, O. (1982). Predictive value of, and observer variability in, several ways of reporting antepartum cardiotocographs. *British Journal of Obstetrics and Gynaecology*, **89**, 434–40.

Freeman, R. K. (1975). The use of oxytocin challenge test for antepartum clinical evaluation of uteroplacental respiratory function. *American Journal of Obstetrics and Gynecology*, **121**, 481–9.

Garite, T. J. and Freeman, R. K. (1979). Antepartum stress test monitoring. *Clinics in Obstetrics and Gynaecology*, **6**, 295–307.

Hage, M. L. (1985). Interpretation of nonstress tests. *American Journal of Obstetrics and Gynecology*, **153**, 490–495.

Krebs, H.-B. and Petres, R. E. (1978). Clinical application of a scoring system for evaluation of antepartum fetal heart rate monitoring. *American Journal of Obstetrics and Gynecology*, **130**, 765–72.

Lavery, J. P. (1982). Nonstress fetal heart rate testing. *Clinical Obstetrics and Gynecology*, **25**, 689–705.

Luthy, D. A., Shy, K. K., Van Bell, G., Larson, E. B., Hughes, J. P., Benedett, T. J., Brown, C. A., Effer, S., King, J. F., and Stenchever, M. A. (1987). A randomised trial of electronic fetal monitoring in premature labor. *Obstetrics and Gynecology*, **69**, 687–95.

Lofgren, O. (1981). Antenatal fetal heart rate testing in high risk pregnancies. *Obstetrics and Gynecology*, **58**, 438–43.

Manning, F. A., Baskett, T. F., Morrison, I., and Lange, I. (1981). Fetal biophysical profile scoring. *American Journal of Obstetrics and Gynecology*, **140**, 289–94.

Montan, S., Olofsson, P., and Solum, T. (1985). Classification of the nonstress test and fetal outcome in 1056 pregnancies. *Acta Obstetricia et Gynecologica Scandinavica*, **64**, 639–44.

O'Leary, J., Andrinopoulos, G. C., and Giordano, P. C. (1980). Variable decelerations and the nonstress test: An indication of cord compromise. *American Journal of Obstetrics and Gynecology*, **137**, 704–6.

Patkos, P., Boucher, M., Broussaid, P. M., Phelan, P., and Platt, L. D. (1986). Factors reflecting nonstress test results in multiple gestations. *American Journal of Obstetrics and Gynecology*, **154**, 1107–8.

Rochard, F., Schifrin, B. S., Goupil, F., Legrand, H., Blottiere, J., and Sureau, C. (1976). Nonstressed fetal heart rate monitoring in the antepartum period. *American Journal of Obstetrics and Gynecology*, **126**, 699–706.

Solum, T. (1980). Antenatal cardiotocography. *Acta Obstetricia et Gynecologica Scandinavica*, **96**, suppl. 1–31.

Solum, T., Ingemarsson, I., and Sjoberg, N.-O. (1979). Selection criteria for antenatal cardiotocography. *Zeitschrift für Geburtshilfe und Perinatologie*, **183**, 212–17.

Solum, T., Ingemarsson, I., and Sjoberg, N.-O. (1980). Selection criteria for oxytocin stress test. *Zeitschrift für Geburtshilfe und Perinatologie*, **184**, 200-4.

Trimbos, J. B. and Keirse, M. J. N. C. (1978). Observer variability in assessment of antepartum cardiotocograms. *British Journal of Obstetrics and Gynaecology*, **85**, 900-6.

Trimbos, J. B. and Keirse, M. J. N. C. (1978). Significance of antepartum cardiotocography in normal pregnancy. *British Journal of Obstetrics and Gynaecology*, **85**, 907-13.

Visser, G. H. A. (1984). Antenatal cardiotocography in the evaluation of fetal well-being. *Australia and New Zealand Journal of Obstetrics and Gynaecology*, **24**, 80-4.

Visser, G. H. A., Redman, C. W. G., Huisjes, H. J., and Turnbull, A. C. (1980). Non-stressed antepartum heart rate monitoring: Implications of decelerations after spontaneous contractions. *American Journal of Obstetrics and Gynecology*, **138**, 429-35.

Chapter 15

Hon, E. H. (1959). The fetal heart rate patterns preceding death in-utero. *American Journal of Obstetrics and Gynecology*, **78**, 47-56.

LaSala, A. P. and Strassner, H. T. (1986). Fetal death. *Clinical Obstetrics and Gynecology*, **29**, 95-103.

Lavery, J. P. (1982). Nonstress fetal heart rate testing. *Clinical Obstetrics and Gynecology*, **25**, 689-705.

MacDonald, D., Grant, A., Sheridan-Pereira, M., Boylan, P., and Chalmers, I. (1985). The Dublin randomized controlled trial of intrapartum fetal heart rate monitoring. *American Journal of Obstetrics and Gynecology*, **152**, 524-39.

Solum, T. and Sjoberg, N.-O. (1980). Antenatal cardiotocography and intrauterine death. *Acta Obstetricia et Gynecologica Scandinavica*, **59**, 481-7.

Westgren, M., Ingemarsson, E., Ingemarsson, I., and Solum, T. (1980). Intrapartum electronic fetal monitoring in low-risk pregnancies. *Obstetrics and Gynecology*, **56**, 301-4.

Chapter 16

Epstein, H., Waxman, A., Gleicher, N. O., and Laueren, N. H. (1982). Meperidine induced sinusoidal fetal heart-rate pattern and reversal with naloxone. *Obstetrics and Gynecology*, **59**, suppl. 22S-25S.

Ingemarsson, I., Liedholm, H., Montan, S., Westgren, M., and Melander, A. (1984). Fetal heart rate during treatment of maternal hypertension with beta-adrenergic antagonists. *Acta Obstetricia et Gynecologica Scandinavica*, **118**, suppl., 95-7.

Jouppila, P., Jouppila, R., Kaar, K., and Merila, M. (1977). Fetal heart rate patterns and uterine activity after segmental epidural analgesia. *British Journal of Obstetrics and Gynaecology*, **84**, 481-6.

Kariniemi, V. and Ammala, P. (1981). Effects of intramuscular pethidine on fetal heart rate variability during labour. *British Journal of Obstetrics and Gynaecology*, **88**, 718–20.

Kariniemi, V., Paatero, H., and Ammala, P. (1981). The effects of oxytocin of fetal heart rate variability during labor. *Journal of Perinatal Medicine*, **9**, 251–4.

Miller, F. C., Quesnel, G., Petrie, R. H., Paul, R. H., and Hon, E. H. (1978). The effects of paracervical block on uterine activity and beat-to-beat variability of the fetal heart rate. *American Journal of Obstetrics and Gynecology*, **130**, 284–8.

Montan, S. and Ingemarsson, I. (1989). Intrapartal FHR-patterns in pregnancies complicated by hypertension. *American Journal of Obstetrics and Gynecology*, **160**, 282–8.

Montan, S., Solum, T., and Sjoberg, N.-O. (1984). Influence of the beta-adrenoceptor blocker atenolol on antenatal cardiotocography. *Acta Obstetricia et Gynecologica Scandinavica*, **118**, suppl., 99–102.

Myers, R. E. and Myers, S. E. (1979). Use of sedative, analgesic, and anesthetic drugs during labor and delivery: Bane or boon? *American Journal of Obstetrics and Gynecology*, **133**, 83–104.

Petrie, R., Yeh, S.-Y., Murata, Y., Paul, R. H., Hon, E. H., Barron, B. A., and Johnson, R. J. (1978). The effect of drugs on fetal heart rate variability. *American Journal of Obstetrics and Gynecology*, **130**, 294–9.

Schifrin, B. S. (1972). Fetal heart rate patterns following epidural anaesthesia and oxytocin infusion during labour. *Journal of Obstetrics and Gynaecology of the British Commonwealth*, **79**, 332–9.

Spencer, J. A. D., Koutsoukis, M., and Lee, A. (1991). Fetal heart rate and neonatal condition related to epidural analgesia in women reaching the second stage of labour. *European Journal of Obstetrics, Gynecology and Reproductive Biology*, **41**, 173–8.

Chapter 17

Arulkumaran, S., Ingemarsson, I., and Ratnam, S. S. (1987). Fetal heart rate response to scalp stimulation as a test of fetal well-being in labour. *Asia-Oceania Journal of Obstetrics and Gynaecology*, **13**, 131–5.

Arulkumaran, S., Michaelson, J., Ingemarsson, I., and Ratnam, S. S. (1987). Obstetric outcome of patients with a previous episode of spurious labour. *American Journal of Obstetrics and Gynecology*, **157**, 17–20.

Clark, S. L., Gimovsky, M. L., and Miller, F. C. (1982). Fetal heart rate response to scalp blood sampling. *American Journal of Obstetrics and Gynecology*, **144**, 706–8.

Clark, S. L., Gimowsky, M. L., and Miller, F. C. (1984). The scalp stimulation test: a clinical alternative to fetal scalp blood sampling. *American Journal of Obstetrics and Gynecology*, **148**, 274–7.

Goodlin, R. and Schmidt, W. (1972). Human fetal arousal levels as indicated by heart rate recordings. *American Journal of Obstetrics and Gynecology*, **114**, 613–21.

Ingemarsson, I. and Arulkumaran, S. (1989). Reactive FHR response to sound stimulation in fetuses with low scalp blood pH. *British Journal of Obstetrics and Gynaecology*, **96**, 562–5.

Ingemarsson, I., Arulkumaran, S., Ingemarsson, E., TambyRaja, R., and Ratnam, S. S. (1986). Admission test. A screening test for fetal distress in labor. *Obstetrics and Gynecology*, **68**, 800–6.
Ingemarsson, I., Arulkumaran, S., Paul, R. H., Ingemarsson, E., TambyRaja, R. L., and Ratnam, S. S. (1988). Fetal acoustic stimulation in early labor in patients screened with the admission test. *American Journal of Obstetrics and Gynecology*, **158**, 70–74.
Krebs, H. B., Petres, R. E., Dunn, L. J., Jordaan, H. V. F., and Segreti, A. (1979a). Intrapartum fetal heart rate monitoring. I. Classification and prognosis of fetal heart rate patterns. *American Journal of Obstetrics and Gynecology*, **13**, 762–72.
MacDonald, D., Grant, A., Sheridan-Pereira, M., Boylan, P., and Chalmers, I. (1985). The Dublin randomized controlled trial of intrapartum fetal heart rate monitoring. *American Journal of Obstetrics and Gynecology*, **152**, 524–39.
Ohel, G., Birkenfeld, A., Rabinowitz, R., and Sadovsky, E. (1986). Fetal response to vibratory acoustic stimulation in periods of low heart rate reactivity and low activity. *American Journal of Obstetrics and Gynecology*, **154**, 619–21.
Serafini, P., Lindsay, B. J., Nagey, D., Pupkin, M., Tseng, P., and Crenshaw, C. (1984). Antepartum fetal heart rate response to sound stimulation: The acoustic stimulation test. *American Journal of Obstetrics and Gynecology*, **148**, 41–5.
Sontag, L. W. and Wallace, R. F. (1936). Changes in the rate of the human fetal heart in response to vibratory stimuli. *American Journal of Diseases in Childhood*, **51**, 583–9.
Spencer, J. A. D. (1991). Predictive value of a fetal heart rate acceleration at the time of fetal blood sampling in labour. *Journal of Perinatal Medicine*, **19**, 207–15.
Visser, G. H. A., Zeelenberg, H. J., de Vries, J. I. P., and Dawes, G. S. (1983). External physical stimulation of the human fetus during episodes of low heart rate variation. *American Journal of Obstetrics and Gynecology*, **145**, 579–84.
Walker, D., Grimwade, J., and Wood, C. (1971). The intrauterine noise: A component of the fetal environment. *American Journal of Obstetrics and Gynecology*, **109**, 91–5.
Westgren, M., Almstrom, H., Nyman, M., and Ulmsten, U. (1987). Maternal perception of sound provoked fetal movements as a measure of fetal well-being. *British Journal of Obstetrics and Gynaecology*, **94**, 523–7.
Yao, Q. W., Jakobsson, J., Nyman, M., Rabaeus, H., Till, O., and Westgren, M. (1990). Fetal responses to different intensity levels of vibroacoustic stimulation. *Obstetrics and Gynecology*, **75**, 206–9.

Chapter 18

Arulkumaran, S. and Ingemarsson, I. (1990). Appropriate technology in intrapartum surveillance. In *Progress in obstetrics and gynaecology*, Vol. 8. (ed. J. W. W. Studd), pp. 127–40. Churchill Livingstone, London.
Baumgarten, K. (1981). The advantages and risks of feto-maternal monitoring. *Journal of Perinatal Medicine*, **9**, 257–74.
Blair, E. and Stanley, F. J. (1988). Intrapartum asphyxia: A rare cause of cerebral palsy. *Journal of Pediatrics*, **112**, 515–19.

Cibils, L. A. (1976). Clinical significance of fetal heart rate patterns during labor. I. Baseline patterns. *American Journal of Obstetrics and Gynecology*, **125**, 290–305.

Dawes, G. S., Moulden, M., and Redman, C. W. G. (1991). System 8000: computerized antenatal FHR analysis. *Journal of Perinatal Medicine*, **19**, 47–57.

Dennis, J., Johnson, A., Mutch, L., Yudkin, M. A., and Johnson, P. (1989). Acid–base status at birth and neurodevelopmental outcome after four and one-half years. *American Journal of Obstetrics and Gynecology*, **161**, 213–20.

Eskes, T. K. A. B., Ingemarsson, I., Pardi, G., Nijhuis, J. G., and Ruth, V. (1990). Consensus statements round table 'Fetal and neonatal distress' (XIIth European Congress of Perinatal Medicine). *Journal of Perinatal Medicine*, **19**, suppl., 126–33.

Fee, S. C., Malee, K., Deddish, R., Minogue, J. P., and Socol, M. L. (1990). Severe acidosis and subsequent neurologic status. *American Journal of Obstetrics and Gynecology*, **162**, 802–6.

Fleischer, A., Schulman, H., Jagani, N., and Randolph, G. (1982). The development of fetal acidosis in the presence of an abnormal fetal heart rate tracing. I. The average for gestational age fetus. *American Journal of Obstetrics and Gynecology*, **144**, 55–60.

Goldaber, K. G., Gilstrap, L. C., Leveno, K. J., Dax, J. S., and McIntire, D. D. (1991). Pathologic fetal acidaemia. *Obstetrics and Gynecology*, **78**, 1103–7.

Hagberg, B. and Hagberg, G. (1984). Prenatal and perinatal risk factors in a survey of 681 Swedish cases. In *Epidemiology of the cerebral palsies* (ed. F. Stanley and E. Alberman), pp. 116–34. Blackwell Scientific, Oxford.

Hall, D. M. B. (1989). Birth asphyxia and cerebral palsy. *British Medical Journal*, **299**, 279–82.

Heden, L., Ingemarsson, I., Ahlstrom, H., and Solum, T. (1991). Induction of labor versus conservative management in prolonged pregnancy: Controlled study. *International Journal of Feto-Maternal Medicine*, **4**, 148–53.

Hobel, C., Hyvarinen, M., and Okada, D. (1973). Prenatal and intrapartum high-risk screening. *American Journal of Obstetrics and Gynecology*, **117**, 1–9.

Hon, E. H. (1962). Electronic evaluation of the fetal heart rate. VI. Fetal distress—a working hypothesis. *American Journal of Obstetrics and Gynecology*, **83**, 333–53.

Ingemarsson, I. and Herbst, A. (1991). The role of electronic fetal monitoring in labour. *Journal of Perinatal Medicine*, **19**, suppl. 1, 134–8.

Ingemarsson, E., Ingemarsson, I., Solum, T., and Westgren, M. (1980). A one-year study of routine fetal monitoring during the first stage of labour. *Acta Obstetricia et Gynecologica Scandinavica*, **59**, 297–300.

Ingemarsson, E., Ingemarsson, I., and Svenningsen, N. (1981). Impact of routine fetal monitoring during labor on fetal outcome with long-term follow-up. *American Journal of Obstetrics and Gynecology*, **141**, 29–38.

Krebs, H. B., Petres, R. E., Dunn, L. J., Jordaan, H. V. F., and Segreti, A. (1979a). Intrapartum fetal heart rate monitoring. I. Classification and prognosis of fetal heart rate patterns. *American Journal of Obstetrics and Gynecology*, **133**, 762–72.

Krebs, H. B., Petres, R. E., Dunn, L. J., Jordaan, H. V. F., and Segreti, A. (1979b). Intrapartum fetal heart rate monitoring. II. Multifactorial analysis of intrapartum fetal heart rate tracings. *American Journal of Obstetrics and Gynecology*, **133**, 773–80.

Leveno, K. J., Cunningham, F. G., Nelson, S., Roark, M., Williams, M. L., Guzick, D., Dowling, S., Rosenfeld, C. R., and Buckley, A. (1986). A prospective comparison of selective and universal monitoring in 34,995 pregnancies. *New England Journal of Medicine*, **315**, 615-19.

Low, J. A., Boston, R. W., and Pancham, S. R. (1971). The role of fetal heart rate patterns in the recognition of fetal asphyxia with metabolic acidosis. *American Journal of Obstetrics and Gynecology*, **109**, 922-9.

Low, J. A., Pancham, S. R., and Worthington, D. N. (1977). Intrapartum fetal heart rate profiles with and without fetal asphyxia. *American Journal of Obstetrics and Gynecology*, **127**, 729-33.

Low, J. A., Galbraith, R. S., Muir, D. W., Killen, H. L., Pater, E. A., and Karchmar, E. V. (1983). Intrapartum fetal hypoxia: A study of long-term morbidity. *American Journal of Obstetrics and Gynecology*, **145**, 129-34.

MacDonald, D., Grant, A., Sheridan-Pereira, M., Boylan, P., and Chalmers, I. (1985). The Dublin randomized controlled trial of intrapartum fetal heart rate monitoring. *American Journal of Obstetrics and Gynecology*, **152**, 524-39.

Murphy, K. W., Johnson, P., Moorcroft, J., Pattinson, R., Russel, V., and Turnbull, A. C. (1990). Birth asphyxia and the intrapartum cardiotocograph. *British Journal of Obstetrics and Gynaecology*, **97**, 470-9.

Nelson, K. B. and Ellenberg, J. H. (1986). Antecedents of cerebral palsy. *New England Journal of Medicine*, **315**, 81-6.

Olofsson, P., Ingemarsson, I., and Solum, T. (1986). Fetal distress during labor in diabetic women. *British Journal of Obstetrics and Gynaecology*, **93**, 1067-71.

Prentice, A. and Lind, T. (1987). Fetal heart rate monitoring during labour—too frequent intervention, too little benefit? *Lancet*, **ii**, 1375-7.

Ruth, V. J. and Raivio, K. O. (1988). Perinatal brain damage: Predictive value of metabolic acidosis and the Apgar score. *British Medical Journal*, **297**, 24-7.

Shields, J. R. and Schifrin, S. (1988). Perinatal antecedents of cerebral palsy. *Obstetrics and Gynecology*, **71**, 899-905.

Spencer, J. A. D. (1991). Monitoring in labour. *Journal of Obstetrics and Gynaecology*, **11**, suppl. 1, 16-19.

Street, P., Dawes, G. S., Moulden, M., and Redman, C. W. G. (1991). Short-term variation in abnormal antenatal fetal heart rate records. *American Journal of Obstetrics and Gynecology*, **165**, 515-23.

Westgren, M., Ingemarsson, E., Ingemarsson, I., and Solum, T. (1980). Intrapartum electronic fetal monitoring in low-risk pregnancies. *Obstetrics and Gynecology*, **56**, 301-4.

Wilson, R. W. and Schifrin, B. S. (1980). Is any pregnancy low risk? *Obstetrics and Gynecology*, **55**, 653-6.

Winkler, C. L., Hauth, J. C., Tucker, M. J., Owen, J., and Brumfield, C. G. (1991). Neonatal complications as related to the degree of umbilical artery acidemia. *American Journal of Obstetrics and Gynecology*, **164**, 637-41.

Index

Note: Figure captions and tables are indicated by *italic page numbers*; abbreviations used: CTG = cardiotocograph; FHR = fetal heart rate

abdominal electrocardiography 21, *25*
abnormal uterine activity 49
   antepartum CTG affected by 247, *248*
   prolonged decelerations associated with *41*, 193–5, 247, *248*
abruptio placentae 49, *51*, *52*
accelerations of FHR 113–24
   in antepartum CTG *113*, 224
   clinical importance of 122–4
   definition of 113
   example traces *6*, *86–7*, *113–23*, *216–17*, *283*
   interruption by variable decelerations 118, *121*, *156*, 156, 162, *164*, 218
   mechanism of 113–21, 153
   periodic accelerations 92, 115, 117, *119*, *121*, *123*
   in preterm fetuses 214
   sporadic accelerations 92, *113*, 115, *118*, *119–21*
acidaemia 28–9, 290–1
   cord compression causing 157–8, 291
acid–base balance
   assessment at birth 204, 206
   measurement of 30
   typical values *31*, 31
   umbilical cord values *31*, 33, 204
acidosis 28–9
   in preterm labour 220, *221*
   role of fetal scalp blood pH sampling 30–2
acoustic stimulation, use with Admission Test *284*, *285*, 285
active state (of fetus) 68
activity, *see* fetal . . . ; maternal activity
Admission Test 274–85
   abnormal/ominous result 276, *277*
      relationship with fetal outcome *278*, *281*
   clinical use of 277–81, 297
   equivocal/suspicious result 275, *276*
      relationship with fetal outcome *278*, *281*
   example traces *15*, *112*, *275–85*
   interpretation of 275–7

normal/reactive result 275, *276*
   relationship with fetal outcome 277, *278*, *281*
   principles of 274
   relationship with fetal distress 277–8, *278*
   vibroacoustic stimulation used in conjunction with 62–3, 283–5
Alexandria units 48
alkalosis 32, *97*
amniocentesis, fetal haemorrhage during 93
amniofusion 177
amniotomy
   effects of 10, *48*
   requirement in direct monitoring 10, 20
   unnecessary in external monitoring 21
anaesthesics
   FHR affected by *78*, 81
   see also local anaesthesics
analgesics, FHR affected by 99, *102*, 269–73
antenatal cardiotocography 3–4, 214–20, 223–54
antepartum CTG
   abnormal uterine activity affecting 61, 247, *248*
   accelerations in *113*, 224
   classification of 233–6
   clinical use of 240–2, 257, 259
   descriptive features of *94*, 223–32
   duration necessary 240–1
   interpretation guidelines for 233
   objective analysis of 237–40
   preterm fetuses 214
   recording procedure for 232–3
   repeating of test 241–2
   technical problems of 243–7
antepartum stillbirths 255
antihypertensive drugs, FHR affected by *196*, 197, 226, 228
Apgar score 203–4
   correlation with CTG changes 287–90
   disadvantage of 204
arrhythmias, FHR affected by 13, *14*, *15*, 94, 105, 107–12

Index

artefacts
  fetal heart rate monitoring *18*, 18, *19*, 83, 229
  uterine activity monitoring 43, *44*
artificial rupture of membranes, FHR response to *126*, *136*, *151*, *197–8*
asphyxia, neurodevelopmental outcome of 291, *293*, 294
atrial flutter 60, 94, 105, 107, *109–10*
atropine, effect of 99, 177
auscultation
  compared with continuous FHR monitoring 1, *2*, *3*, 255
  use when continuous FHR record unsatisfactory 14
autocorrelation, signal-to-noise ratio improved by 18, 21
autonomic nervous system
  baseline FHR influenced by 53–4
  pharmacological blockade of, FHR variability affected by 54, 56–7
AV (atrioventricular) block *61*, 107, *111–12*
averaging process 19
  effect of 19–20
awake state 69

background noise (in uterus) 283
baroreceptors 53, 154
baseline fetal heart rate 83–112
  in antepartum CTG 225–7
  autonomic nervous system control of 53–4
  difficulties of estimation 14, 83, *84*
  estimation between contractions *83*, 83
  factors affecting 53–4
  meaning of term 83
  normal range 4, 84–6
    correlation with low Apgar scores 287–8
    incidence during labour 5, 287
  preterm fetuses 214–18
  see also bradycardia; tachycardia
baseline FHR variability, see variability of fetal heart rate
behavioural states (of fetus)
  age when discernible 69–70
  diurnal pattern 70–1
  effect of external manipulation *62*, *72*, *73–4*, *282*, 282
  FHR affected by 7, 8, 67–74
  tachycardia associated with *9*, 91
β-receptor agonists
  FHR affected by 14–15, *16*, *98*, 99, *201–2*, 267
  side-effects 202
  see also terbutaline

β-receptor blockers, FHR affected by 226, 228, 268–9
biphasic decelerations *166*, 168, 183
  see also combined decelerations
birth asphyxia, effects of 291, *293*, 294
bradycardia 99–109
  in antepartum CTG 226
  correlation with low Apgar scores 86, *287–8*, *289*
  definition 5, 85
  delivery necessary if continuing after low FHR variability *103–4*, 205
  drug-induced 99, *101*, *265*, 265–6
  example traces *6*, *15*, *22–3*, *47*, *61*, *86–8*, *100–7*, *205*, *209–11*, *231*, *250*, *260–1*, *265*, *279*, *284–5*
  fetal origin 99–101
  incidence during labour *5*, 85, 221, 287
  maternal origin 101–9
  physiological effects associated with 85–6, *104*, *106*
  in terminal FHR record 86, 260, *260–1*
brain damage, asphyxia associated with 291, *293*, 294
Braxton Hicks contractions 36
  FHR response to 28, 35, *36*, *61*, 229, *261*
breech presentation
  bradycardia associated with *103*
  tachycardia associated with *91–2*
  variable decelerations associated with *177*, 179, *180–1*

Caesarean section
  anaesthesia for, CTG trace during *78*
  indications for *15*, *23*, *62–3*, *66*, *139*, 145, 150, *165*, *188–9*, *229*, *246*, *250*, *253*, *265*, *279*, *280*
  rates 7, 298, 300
    preterm babies 213
  rupture of scar from previous section *47*
  unnecessary 3, 7, 301
cardiotocograph, see CTG
catheter-tip transducer 43
  uterine activity trace using *44*
chemoreceptors 53
chest-and-knees position, FHR affected by 77, 78, *175*
chorioamnionitis *262*
clinical considerations 286–301
combined decelerations 182–92
  aetiology 183–91
  correlation with low Apgar scores 287–8
  example traces *77*, *92*, *103*, *106*, *127*, *182–3*, *185–92*, *208*, *221*
  incidence during labour 5, 183–4, 287
  uterine activity associated with *184*

Index

compensatory tachycardia 87, 89, *158*, *159*, *161*, *165*, *169*, *196*
consumer attitudes/decisions 4, 298, 301
continuous fetal heart rate (FHR) monitoring
  advantages 1-4
  compared with intermittent monitoring 1-11, 296-8
  disadvantages 4-10
  intrapartum stillbirth incidence affected by 255, 286
  side-effects 10
contractions, *see* uterine contractions
cord arterial blood
  acid-base balance
    assessment at birth 204, 206
    typical values *31*, 33
  characteristics *31*, 31, *32*, 33, 204
cord compression
  acidaemia associated with 157-8, 291
  FHR decelerations caused by 118, *121*, 153-6, *195*, 195, *259*
  relief management measures 174-7
cord dips 153, *154*, *259*
  see also variable decelerations
cord prolapse 175-7
  FHR affected by *176*, *177*, *279*
CTG (cardiotocograph) 2-3, 12
  incidence of changes during labour 5, 287
  poor interpretation of CTG traces 7

Dawes-Redman criteria, antepartum CTGs analysed using 237
dead fetus, maternal heart rate recorded via 16, 16-17, *17*
decelerations of FHR
  in antepartum CTG 228-32
  characteristics of 128-30
  classification of 125-8
  combined decelerations 182-92
    aetiology 183-91
    correlation with low Apgar scores 287-8
    example traces 77, 92, *103*, *106*, *127*, *182-3*, *185-92*, *208*, *221*
    incidence during labour 5, 183-4, 287
  definition of 125
  early decelerations 130-2
    correlation with low Apgar scores 287-8, 289
    example traces *96*, *130-4*, *276*
    incidence during labour 5, 221, 287
  incidence of various types 5, 221, 287
  late decelerations 133-51
    in antepartum CTG *229*
    causes 133, 135, 137
    clinical importance 137-52

    correlation with low Apgar scores 287-8, 289
    example traces 3, 4, 9, 22-3, 85, 90, 92-3, 125-6, *129*, *134-6*, *138-52*, *210*, *226*, *229*, *250*, *253*, *257*, *261*, *266*, *277-8*
    incidence during labour 5, 221, 287
  preterm fetuses 214, *216-20*
  prolonged decelerations 193-202
    aetiology 193-9
    definition 193
    example traces *41*, *52*, *127*, *135*, *139*, *148*, *150-1*, *194-8*, *200-1*, *247*, *267*, *280*
    management measures 199-202
  sporadic decelerations 220, *263*
  uniform decelerations
    characteristics 128, *129*
    timing of 130-52
  variable decelerations 153-81
    aetiology 153-7, 180
    in antepartum CTG *229*, *230-1*
    characteristics 128, 130, 153
    correlation with low Apgar scores 287-8, 289
    example traces 5, 6, 13, 76-7, 84, 92, 95, 97, *100*, *103*, *106*, *121*, *130-1*, *154-78*, *180-1*, *208-9*, *221*, *229-30*, *232*, *259*, *262*, *290*
    incidence during labour 5, 153, 221, 287
    in terminal FHR record *262*, 263
  see also combined . . . ; early . . . ; late . . . ; prolonged . . . ; uniform . . . ; variable decelerations
depression of baby at birth
  factors causing 10, 47, *103*, 203-4
  scoring systems for 203-4, 206
diabetes mellitus, effects of *227*, 296
direct fetal electrocardiography 20
direct monitoring of FHR, *see* internal monitoring (of) fetal heart rate
Doppler ultrasound technique 21
doubling of FHR signal 17-18, *205*
drugs, FHR affected by 56-7, 99, 265-73
duration of labour, effect on fetal distress 279, 281

early decelerations 130-2
  correlation with low Apgar scores 287-8
  example traces *96*, *130-4*, *276*
  incidence during labour 5, 221, 287
  relationship with uterine contraction *132*
epidural analgesia
  FHR affected by 99, *101*, 137, 266, 266-7, *267*
  length of labour affected by 203

epidural analgesia (*cont.*)
  maternal hypotension caused by 196, 266
  maternal pyrexia associated with 98, 267
epileptic fits, FHR changes associated with
  195, *196*, 211
external monitoring
  fetal heart rate 12, 21–2
    compared with direct monitoring 21,
    26, 79, 205
    effect of repositioning of transducer *19*
    example traces *4, 5, 6, 8, 9, 25–6,
    50–1, 79, 171, 205, 282, 293*
  uterine activity 3, 43
    example traces *3, 4, 5, 6, 8, 9, 22–5,
    36, 42, 50–1, 84, 86–8, 95–6, 146,
    158, 167, 257, 282*
extrasystolies, FHR record affected by 13,
  13, 108, 109

false-positive interpretation (of FHR) 4–7
  means of reducing 74
fetal activity
  FHR changes due to 113–15, *113–21,
  176, 214–17*
  preterm fetus 214
fetal bleeding, effects of 93, *197*, 197
fetal breathing movements, FHR affected by 74
fetal head compression, FHR affected by
  99, *100*, 178–9
fetal heart rate, *see* baseline fetal heart rate
fetal heart rate variability, *see* variability of
  fetal heart rate
fetal hypoxia 27–9
  FHR affected by 28, 81, 87, *89–92*, 101,
  103–4, 151, 164–6
fetal quiescence
  CTG traces showing *8, 68, 69*
  normal length of period *8*, 72, 74, 224
fetal scalp blood
  acid–base balance values *31*, 31
  sampling technique 32–3, 197
fetal scalp blood pH
  clinical interpretation of values 29–30,
  295
  deceleration classification and 159, 199,
  201
  measurement error range 30
  normal range 31–2
  role of sampling 30–3, *143*, 177
fetal vagal reflex 198–9
fever, tachycardia caused by *95*, 98
FIGO (International Federation of
  Gynaecologists and Obstetricians)
  recommendations
  criteria for monitoring 294
  normal range for baseline FHR 85

Fischer score (for antepartum CTG) *233*,
  233–4

halving of FHR signal *17*, 18
heart malformations
  baseline FHR affected by 105–10
  FHR variability affected by 14, 60, *61*, 81
hiccups
  convulsions mistaken for *293*
  FHR affected by *259*
high-risk groups 236, 268, 295–6
  see also diabetes . . . ; hypertension . . . ;
  preterm . . .
hormones, FHR variability affected by 54,
  56–7
hyperstimulation of uterus 40, 40–1, 184
  effect of oxytocin 40, 45, 47, 52, 84, *106*,
  142, 184, *190*, 191, 193, *250*
  reduction by β-receptor agonists 40, *40*,
  51, 52, 98, *104*, 193, *194*, 199–201,
  250
hypertension
  drug treatment of *196*, 197, 226, 228
  FHR affected by *186–7*, 228, *259*
  ominous traces associated with 268, 269,
  296
hyperventilation, fetal blood pH affected by
  32, 97
hypoactivity 49
  management options 49, 178
hypotension, FHR changes associated with
  105, *107*, 137, 195–6
hypovolaemia, FHR changes associated with
  102
hypoxaemia, FHR changes due to *9*, 114,
  125
hypoxia 27–9
  acidosis due to 28–9
  causes of 27–8
  fetal response to 28
  FHR affected by 28, 81, 87, *89–92*, 101,
  103–4, 151, 164–6

idiopathic bradycardia 101, *105*
intermittent auscultation 1, 298, 301
  compared with continuous FHR recording
  1, *2, 3*
intermittent monitoring 297–8
  advantages 298
  compared with continuous FHR recording
  298
internal monitoring
  fetal heart rate 12
    compared with external monitoring 21,
    26, 79, 205

Index

example traces *2, 6, 9, 13–17, 24–6,*
  41, 47, 73–7, 79, 112, 125, 131, 185,
  205, 256–7
  risks involved 300
  technical aspects 12–26
  uterine activity 2, 43–5
    example traces *2, 6, 9, 35, 39–42, 44,*
      45, 46–7, 52, 73–7, 85, 97, 106–7,
      125, 131–2, 140, 185, 190–2, 194
    indications for 44–5
intervention, unnecessary 3, 7, 301
intervillous perfusion, effect of contractions
  on 34–5
intrapartum CTG scoring system *299,* 299
intrapartum stillbirths
  effect of continuous FHR monitoring
    255, 286
  incidence of 255, 286
intrauterine growth retardation *221, 229,*
  295–6
intrauterine monitoring, *see* internal
  monitoring of uterine activity

jitter of FHR record *18,* 18, *229*

knees-and-chest position, FHR affected by
  77, 78, *175*
Krebs score (for intrapartum CTG) *299,* 299

labour
  CTG scoring systems during 299
  duration of 279, 281
late decelerations 133–51
  causes 133, 135, 137
  classification 133
  correlation with low Apgar scores *287–8*
  example traces *3, 4, 9, 22–3, 85, 90,*
    92–3, 125–6, 129, 134–6, 138–52,
    210, 226, 229, 250, 253, 256–7, 261,
    266, 277–8
  factors affecting 138
  incidence during labour 5, *221,* 287
  management measures 141, 144–5, 150
  relationship with uterine contraction *132*
local anaesthetics
  FHR affected by 99, *101, 157, 265,* 265–6
  see also epidural . . .

manual manipulation, FHR response to *62,*
  72, 73–4, 224, *282,* 282
maternal activity, FHR changes caused by
  41, 198, 199
maternal fever, tachycardia caused by *95,* 98

maternal heart rate, confusion with fetal
  rate 14–17, 22, *26,* 226
maternal hypertension
  drug treatment for *196,* 197, 226, 228
  FHR affected by *186–7, 259*
  ominous FHR changes associated with
    268, 269, 296
maternal hypotension, FHR changes
  associated with 105, *107,* 137, 195–6
maternal position
  FHR affected by 77, 78, 105, *107, 136,*
    174–5, *175, 231*
  FHR recording not affected by 298
  uterine activity affected by *39,* 42–3
maternal seizures, FHR changes associated
  with 195, *196*
medicalization of childbirth 300–1
monitoring of FHR
  complications due to 300
  consumer attitudes/decisions 4, 298, 301
  continuous monitoring 1–11
    advantages 1–4
    disadvantages 4–10
  decision flow-chart for *297*
  FIGO guidelines 294
  intermittent monitoring 297–8
  recommendations in clinical practice
    294–301
  risk assessment criteria 295–6
  risks involved 300–1
Montevideo units 45–6
  calculation for 46, *48*
  typical values 46
morbidity measures 286
mortality measures 286

narcotic analgesics
  FHR affected by 99, *102,* 269–73
  neonatal condition affected by 204
neonatal acidaemia 290–1
neonatal condition, assessment of 203–6
neonatal outcome measures 287–91
non-stress test (NST) 223
  see also antepartum CTG

occiput posterior position
  FHR patterns associated with *100, 178,*
    178–9
  incidence during labour 178
OCT (oxytocin challenge test) 35, 250–3
  contraindications 252
  drawbacks 252, 254
  example traces *250–3*
  interpretation 251–3
  procedure 251

oligohydramnios   177, *218*, 228, 232, 296
ominous CTG
   correlation with condition at birth   8-9, 277-8
   definition of   *234*, 276-7
   examples of   *22-3*, *61*, *187*, *236*, *277*, *278*
   hypertensive women   268, 269, 296
outcome measures   286-94
overstimulation of uterine activity   45, *47*, 52, *106*, 184
Oxford Sonicaid System 8000   237
   advantages of   240
   CTG records with numerical analysis   237-9
oxygen tension   27
oxytocin
   hypoactivity corrected by   49, 178, *185-92*
   overstimulation by   *40*, 45, *47*, 52, *84*, *106*, *142*, 184, *190*, 191, *250*
   response to   45, 49, 101, *104*, *126-7*, *129*, *136*, *140-3*, 191, 267
      importance of monitoring   191
   see also OCT (oxytocin challenge test)

palpation, uterine activity monitored by   42
paper speeds, effect on FHR records   20, 20, 58, *59*, 60
paracervical block, FHR affected by   *157*, 265, 265-6
pathological CTG
   definition of   234
   example traces   *146-7*, *226-7*, *229*, *235*, *236*, *246*
   incidence of   234
periodic accelerations   92, 115, 117, *119*, *121*, *123*
permanent record, advantage of   3
pethidine, FHR affected by   99, 269, *270-3*
placental abruption   23, 49, *51*, *52*, *90*, 275
placental function, antenatal assessment of   3-4
polyhydramnios   48, 49, *151*, 292
post-term pregnancy   296
pre-eclampsia, FHR affected by   90, *220-1*, 227
prematurity
   FHR variability caused by   80
   tachycardia caused by   93-4, *94-5*
preterm FHR patterns   213-22
   clinical aspects   213
   factors influencing   213-22
   technical problems encountered   213
preterm labour
   FHR patterns during   220-2
   inhibition of   15, *16*, 98, 99
prolonged contractions   40-1

prolonged decelerations   193-202
   aetiology   193-9
   definition   193
   example traces   *41*, *52*, *127*, *135*, *139*, *148*, *150-1*, *194-8*, *200-1*, *247*, *267*, *280*
   management measures   199-202
prolonged labour, effects of   *96*, 98
propranolol (β-blocker), FHR affected by   268, 268
prostaglandins, response to   49, *50-1*, *186-7*
pseudosinusoidal variation of FHR pattern   25, 67
pudendal block, FHR affected by   *157*, 265
pulsed Doppler ultrasound technique, compared with direct fetal ECG   21

quiescence periods
   CTG traces showing   *8*, *68*, *69*
   incidence of   71
   length of   *8*, 72, 74, 224
   preterm fetuses   214

reactive Admission Test   275, *276*
reactive antepartum CTG
   criteria for interpretation   233
   definition of   *234*
   example traces   *223*, *225*, *234*
   incidence of   234
reactivity, effect of fetal behavioural states   67-74, 224
rebound tachycardia   *165*, 166
references listed (for this book)   302-17
rhesus isoimmunization, FHR response to   63, *65*, 87

sampling techniques
   fetal scalp blood   32-3
   umbilical cord arterial blood   33, 204
scalp electrode   20
   FHR changes during application   *198*
   FHR record compared with external monitoring   21, *26*, *79*, *205*
   importance of positioning   20, *24-5*
scalp pH values, interpretation of values   29-30
scalp stimulation test   282, *283*
second stage of labour   203-12
   clinical aspects   203-6
   continuous FHR monitoring necessary   298
   duration   203
   interpretation of FHR patterns   206-11
signal loss, causes of   12-18, *171*, *205*
signal-processing techniques   18-24

sinusoidal pattern of FHR variability 60-1, 63-4, *64-7, 263*
slow labour progress
 management options 49, 178, *185*
 see also oxytocin
spontaneous uterine activity
 FHR response to *35*
 see also Braxton Hicks contractions
sporadic accelerations 92, *113*, 115, *118*, *119-21*
sporadic decelerations *220, 263*
stillbirths
 aetiology 256-9
 incidence 255
stimulation (of fetus)
 FHR response to *62-3, 72*, 73-4, *86-7*, *93, 119, 120, 136*, 224, 281-5
 see also manual manipulation; vaginal examination; vibroacoustic stimulation
supine position
 FHR affected by 105, *107, 136, 175, 231*
 uterine activity affected by *39*, 42-3
suspicious (non-reassuring) CTG
 definition of *234*, 275
 example traces *225, 227, 235, 276*
 incidence of 234
System 8000 237
 advantages of 240
 CTG records with numerical analysis *237-9*

tachycardia
 in antepartum CTG 225-6
 in combination with other FHR changes 288-90
 compensatory tachycardia 87, 89, *158*, *159, 161, 165, 169, 196*
 correlation with low Apgar scores *287-8*, 288, *289*, 290
 definition of 5, 84-5
 example traces *60, 85, 89-98, 122, 146*, *152, 208, 211, 225-6, 248, 261, 263*, *283, 292-3*
 fetal origin 87-94, *225*
 incidence during labour 5, *85*, *221*, 287
 with low/absent FHR variability 80-1, *226, 289*, 290, *292*
 maternal origin 98-9
 physiological effects 85
 rebound tachycardia *165*, 166
 in terminal FHR record *261*
 transient episodes *93, 116, 117*
technical aspects of FHR monitoring 12-26
terbutaline, effects of *40, 51, 52*, 98, *104*, 193, *194, 199-201, 250*
terminal FHR patterns 86, 199, *229*, 255-64

tocodynamometry, uterine activity monitored by 43, *146*
tocography 2-3
twin pregnancy *243-6*, 247

ultrasound FHR monitoring 21-2, *26*
umbilical cord arterial blood
 acid-base balance
  assessment at birth 204, 206
  measurement of 33, 204
  relevance of 33
  typical values *31*, 33
 flow during variable deceleration *155*, 156
 pH values *31*, 31, *32*
umbilical cord compression
 FHR changes during 118, *121*, 153-6, *195*, 195, *259*
 see also cord compression
umbilical cord prolapse 175-7, *279*
uniform decelerations
 characteristics 128, *129*
 timing of 130-52
 see also early . . . ; late decelerations
uterine activity 34-52
 abnormal activity 49
 antepartum CTG affected by 247, *248*
 prolonged decelerations associated with *41*, 193-5, 247, *248*
 bradycardia caused by 52, 101, *106*
 hyperstimulation of *40*, 40-1, *47*, *84*
 monitoring of 42-5
 physiology 36-41
 in pregnancy 36
 quantification of 45-9
uterine background noise 283
uterine contractions
 amplitude of 37, *42*, 42
 basal tone of 39, *40-1*, *45*, *52*
 characteristics 37-9
 duration of 38
 during labour 37
 frequency of 38
 intervillous perfusion affected by 34-5
 prolonged contractions 40-1
 recorded with FHR monitoring 1-3
 relationship with uniform decelerations *132*
 shape of tocograph 34, *38*, 38

vaginal bleeding, FHR affected by *90, 197*
vaginal examination, FHR changes in response to *62-3, 86-7, 120, 136*

326　Index

variability of fetal heart rate 53–82
  absent/silent pattern
    in combination with other FHR
      changes 289, 290
    correlation with low Apgar scores
      287–8, 289
    example traces 4, 8, 19, 47, 56, 60–1,
      68–73, 79–81, 89, 92–3, 95, 104, 120,
      123, 146–7, 169, 173, 200–1, 221, 229,
      236, 246, 260–3, 272, 275, 278, 280,
      284
    incidence during labour 5, 221, 287
    meaning of term 5, 54
    in terminal FHR record 260, 260–3
  in antepartum CTG 227–8
  classification of 54–8
  effect of behavioural states 67–74
  estimation between contractions 59, 74
  factors affecting 53–4, 56–7
  incidence of various types 5
  interpretation of 74–82
  long-term variability
    compared with short-term variability
      58–67
    estimation of 54, 55, 227–8
    meaning of term 54
  low-variability pattern
    causes 76, 80–1
    correlation with low Apgar scores
      287–8
    example traces 3, 8, 22–3, 56, 68–71,
      73, 85, 89, 90, 96–7, 120, 122–3, 125,
      134, 139, 142–5, 148–52, 180–1,
      185–9, 196, 205, 208–11, 215, 220,
      226–9, 235, 238–9, 250, 253, 256–7,
      261, 265, 270–3, 276–7, 282, 292
    incidence during labour 5, 221, 287
    interpretation 80–2
    meaning of term 5, 55
  normal pattern
    correlation with low Apgar scores
      287–8
    example traces 2, 5, 6, 57, 71–3, 77,
      86–7, 94, 98, 103, 123, 126–7, 129,
      148, 168, 172, 192, 194, 200–1, 206–8,
      215–18, 225–6, 234, 237, 248, 270–3,
      276, 284, 290
    incidence during labour 5, 287
    meaning of term 5, 54
  preterm fetuses 214
  pseudosinusoidal variation 25, 67

  saltatory (high-variability) pattern
    correlation with low Apgar scores
      287–8, 289
    effect of maternal position 77, 78
    example traces 57, 69, 75–7, 88, 115,
      117, 167, 208
    incidence during labour 5, 221, 287
    interpretation 76–9
    meaning of term 5, 55
  short-term variability
    compared with long-term variability
      58–67
    meaning of term 54
  sinusoidal pattern 60–1, 63–4, 67
    example traces 64–7, 263
    possible causes 63–4
    in terminal FHR pattern 263
variable decelerations 153–81
  in absence of accelerations 161–2, 163, 178
  accelerations interrupted by 118, 121,
    156, 156, 162, 164, 218
  aetiology 153–7, 180
  atypical traces 161–8, 163, 169
  and baseline FHR variability 169–74
  biphasic decelerations 166, 168
  characteristics 128, 130, 153
  correlation with low Apgar scores 287–8
  duration 160, 161
  example traces 5, 6, 13, 76–7, 84, 92, 95,
    97, 100, 103, 106, 121, 130–1, 154–78,
    180–1, 208–9, 221, 229–30, 232, 259,
    262, 290
  incidence during labour 5, 153, 156, 221,
    287
  loss of FHR variability during 166, 167
  lower baseline after 165, 165
  management measures 174–9
  predictive value 159–74
  rebound tachycardia after 165, 166
  shape classification 161–8
  size classification 159–61
  with slow recovery 164, 164
  in terminal FHR record 262, 263
  U-shaped traces 161, 162
  V-shaped traces 161, 162
vena cava syndrome 105, 107, 137
vibroacoustic stimulation
  fetal response to 62–3, 73–4, 93, 119,
    224, 283–5
  possible harmful effects 285
vomiting, FHR deceleration due to 198